CISM International Centre for Mechanical Sciences

Courses and Lectures

Volume 600

For more than 40 years the book series edited by CISM, "International Centre for Mechanical Sciences: Courses and Lectures", has presented groundbreaking developments in mechanics and computational engineering methods. It covers such fields as solid and fluid mechanics, mechanics of materials, micro- and nanomechanics, biomechanics, and mechatronics. The papers are written by international authorities in the field. The books are at graduate level but may include some introductory material.

More information about this series at http://www.springer.com/series/76

Albrecht Bertram · Samuel Forest
Editors

Mechanics of Strain Gradient Materials

Springer

Editors
Albrecht Bertram
Otto-von-Guericke-University Magdeburg
Magdeburg, Germany

Samuel Forest
CNRS
MINES ParisTech
Évry, France

ISSN 0254-1971 ISSN 2309-3706 (electronic)
CISM International Centre for Mechanical Sciences
ISBN 978-3-030-43829-6 ISBN 978-3-030-43830-2 (eBook)
https://doi.org/10.1007/978-3-030-43830-2

This Springer imprint is published by the registered company Springer Nature Switzerland AG
The registered company address is: Gewerbestrasse 11, 6330 Cham, Switzerland

Preface

Many materials show a different elastic or plastic behavior if tested on large or on small samples. Such size effects can be ascribed to internal length scales related to the microstructure. This behavior is for instance observed in indenter tests, torsion, bending, and shear banding. Such effects cannot be simulated by classical constitutive models in which the stresses depend locally on the (dimensionless) strains. One possible extension of the classical models relies on the constitutive inclusion of higher strain gradients, which involve the dimension of length. This allows for an introduction of internal length scales in the constitutive model. Strain gradient models can further be employed for the regularization of singularities in the classical solutions, which makes them also advantageous from a numerical point of view. Moreover, they allow for the conceptually sound introduction of line and point forces into continuum mechanics.

The balance laws and boundary conditions for such materials can be derived by resorting to variational principles. For each primal higher strain gradient, a conjugate stress tensor of the same tensorial order has to be introduced, for which a constitutive law is needed. Material modeling becomes a challenging task for such materials.

In the elastic case, all stress tensors may depend on all of the corresponding strain tensors. Even in the linear case, this leads to an enormous amount of material constants. This can, however, be drastically reduced by the assumption of symmetry properties like isotropy or centro-symmetry.

In the non-linear case of finite deformations, one has to satisfy invariance principles, which is neither trivial. In the case of plastically deforming metals, primal higher strain gradients are chosen with the aim of describing the behavior of geometrically necessary dislocations.

The book is limited to strain gradient modelling and related theories. It does not consider other generalized continuum approaches like Cosserat or micromorphic mechanics for which other books, like Eringen's contributions, are available.

In the course the following topics have been considered:

- Experimental findings for mechanical size effects;
- Balance laws and boundary conditions for strain gradient materials;
- The linear theory of elasticity and plasticity of strain gradient materials;
- The derivation of effective strain gradient materials using asymptotic homogenization methods;
- The application to strain gradient plasticity, crystal plasticity and the associated computational challenges;
- The finite strain gradient theory for large deformations;
- The application of strain gradient models to fracture and damage and to micro-to-macro transitions (media with microstructure).

Berlin, Germany — Albrecht Bertram
Paris, France — Samuel Forest

Contents

Chapter 1
The Experimental Evidence for Higher Gradient Theories

Wolfgang H. Müller

Abstract In this document, it will be attempted to summarize the currently available experimental evidence for the need of higher gradient continuum theories. They allow to capture size effects, e.g., within the elastic behavior of materials with an internal substructure, which gives rise to additional length scale parameters. These are not available in elastic theories of the Cauchy–Boltzmann type. First, some theoretical background information regarding the additional parameters on higher gradient continuum theories will be given. Then the pertinent constitutive equations of second gradient materials for both solids and fluids will be presented, such that they agree with nomenclature used in the experiments. Finally, experimental data showing the size effects for various materials are compiled stemming from different sources. Some own results are also included, and the literature is extensively reviewed.

The Need for Higher Gradient Continuum Theories

Often, the deformation behavior of modern high-performance materials cannot be described by using the traditional Cauchy–Boltzmann continuum. This is because in these materials an additional intrinsic length scale on the microlevel occurs, which is not captured in the traditional approach. From an experimental point of view, this leads to the "so-called" *size effect*, such that the stiffness of a specimen seemingly depends on its size. In order to model such behavior, generalized continuum theories (GCTs) have proven to be most useful. A particular form of GCT are material models that take higher gradients of strain or velocity into account; see Bertram (2019), Dell'Isola et al. (2015). They allow modeling size-dependent stiffnesses, for example, in pantographic structure deformation and miniature beam bending,

---

W. H. Müller (✉)
Chair of Continuum Mechanics and Constitutive Theory, Institute of Mechanics,
Technische Universität Berlin, Einsteinufer 5, 10587 Berlin, Germany
e-mail: wolfgang.h.mueller@tu-berlin.de

A. Bertram and S. Forest (eds.), *Mechanics of Strain Gradient Materials*,
CISM International Centre for Mechanical Sciences 600,
https://doi.org/10.1007/978-3-030-43830-2_1

or size-dependent flow behavior, e.g., in blood rheology. In what follows, we will present the results of various experiments performed in context with higher strain gradient theory.

Theoretical Background

Remarks on various aspects regarding the historical development of higher gradient theories can be found in Dell'Isola et al. (2017). A detailed summary of gradient material theory can be found in Bertram (2019) and in this monograph. The theory of higher gradient materials emphasizes the energetic concept of virtual power, introduces Lagrangians, and uses variational calculus and functional theory. All of this makes it difficult for the engineer to follow. Therefore, as a stimulant, it seems only fair to remind us of the following goal of higher gradient theory. The idea is to capture force and deformation effects associated with the surface, edges, or corners of a material body, $\mathcal{B}$, of volume V and to understand them in terms of strong gradients when entering the bulk. In order to quote from Mindlin (1965): "It seems worthwhile, however, to formulate completely the equations of the simplest theory, incorporating the second gradient of strain, and to exhibit some typical solutions. The cohesive force and surface tension, or surface energy per unit area, which appear, may have some bearing on criteria of failure of solids and on a variety of surface phenomena observed in both solids and liquids."

In the following, some constitutive equations for higher gradient materials are presented. In order to discuss their measurement, the additional material parameters for higher gradient fluids as well as solids are shown.

Some Constitutive Equations for Higher Gradient Fluids

Explicit constitutive equations for the hyperstresses of second-order fluids can be found in Silber et al. (2007)

$$\begin{aligned}
\overset{(2)}{\boldsymbol{\sigma}} &= -p\mathbf{I} + \nu_1 \operatorname{Tr} \boldsymbol{D}\mathbf{I} + \nu_2 \boldsymbol{D}, \\
\overset{(3)}{\boldsymbol{\sigma}} &= \mathbf{I}\left(\nu_3 \nabla \operatorname{Tr} \boldsymbol{D} + \frac{\nu_4}{2}\nabla \cdot \boldsymbol{D}\right) + \frac{\nu_4}{4}\left[(\nabla \operatorname{Tr} \boldsymbol{D})\,\mathbf{I} + \mathbf{I}\,(\nabla \operatorname{Tr} \boldsymbol{D})\right] + \\
&\quad \frac{\nu_5}{2}\left[(\nabla \cdot \boldsymbol{D})\,\mathbf{I} + \mathbf{I}\,(\nabla \cdot \boldsymbol{D})\right] + \nu_6 \boldsymbol{D}\nabla + \frac{\nu_7}{2}\left(\boldsymbol{D}\nabla + \nabla \boldsymbol{D}\right).
\end{aligned} \tag{1.1}$$

Note that in Silber et al. (2007) gradients are applied from the left-hand side. In order to be consistent with the conventions of this book, the representation of the second hyperstress tensor is modified. The various unit tensors of different ranks are explained in the appendix of Silber et al. (2007). ν_i are numerous viscosity coefficients in context with the velocity gradient $\boldsymbol{D}$ or its (non-symmetric) gradient $\nabla\boldsymbol{D} = \frac{1}{2}(\nabla\nabla\boldsymbol{v} + \nabla\boldsymbol{v}\nabla)$. More general forms for nth gradient fluids are presented in Chap. 8 of Bertram (2019).

Some Constitutive Equations for Higher Gradient Solids

The constitutive laws for finite third-order anisotropic gradient elastic solids are studied in Chap. 3 of Bertram (2019). In particular, expressions for the strain energy density are shown leading to more than 1000 stiffness coefficients, which is a theoretician's dream but an experimentalist's nightmare. In a more restricted way, the situation is analyzed in Dell'Isola et al. (2009) for isotropic second-order materials subjected to large deformations. This is of practical use, for example, if the so-called pantographic structures are examined, where a small deformation theory is insufficient. The strain energy is therefore first expressed in terms of the deformation gradient and its (Lagrangian) gradient, $\boldsymbol{F}$ and $\boldsymbol{F}\nabla_X$, respectively. The principles of rational mechanics are then used to arrive at the following form of the (objective) strain energy density w.r.t. the reference placement $\boldsymbol{X}$

$$w = \frac{1}{2}\left(E_{IJ}C_{IJKL}E_{KL} + 2E_{IJ,K}H_{IJKLP}E_{LP} + E_{IJ,K}G_{IJKLPQ}E_{LP,Q}\right), \tag{1.2}$$

where capital indices refer to a representation and differentiation w.r.t. Cartesian coordinates of the reference placement. The (constant) stiffness tensors $\boldsymbol{C}$, $\boldsymbol{H}$, and $\boldsymbol{G}$ are of fourth, fifth, and sixth rank, respectively. They satisfy certain symmetry properties following from the equality of strain energy density and its complement in linear elasticity and from the symmetries of the Green–Lagrange tensor $\boldsymbol{E} = \frac{1}{2}\left(\boldsymbol{F}^\top \cdot \boldsymbol{F} - \mathbf{I}\right)$ and its derivative w.r.t. to reference placement. If the material is isotropic, then $\boldsymbol{H}$ vanishes and $\boldsymbol{C}$ and $\boldsymbol{G}$ can be expressed by sums of products of two and three unit tensors of second rank, respectively. This way $\boldsymbol{C}$ is reduced to two stiffness parameters (the two Lamé coefficients) and $\boldsymbol{G}$ to five

$$\begin{aligned}
C_{IJKL} &= \lambda\delta_{IJ}\delta_{KL} + \mu\left(\delta_{IK}\delta_{JL} + \delta_{IL}\delta_{JK}\right), \quad H_{IJKLP} = 0\,,\\
G_{IJKLPQ} &= c_2\left(\delta_{IJ}\delta_{KL}\delta_{PQ} + \delta_{IJ}\delta_{KP}\delta_{LQ} + \delta_{IK}\delta_{JQ}\delta_{LP} + \delta_{IQ}\delta_{JK}\delta_{LP}\right) +\\
&\quad c_3\delta_{IJ}\delta_{KQ}\delta_{LP} + c_5\big(\delta_{IJ}\delta_{JL}\delta_{PQ} + \delta_{IK}\delta_{JP}\delta_{LQ} +\\
&\qquad\qquad \delta_{IK}\delta_{JQ}\delta_{LP} + \delta_{IL}\delta_{JK}\delta_{LP} + \delta_{IP}\delta_{JK}\delta_{LQ}\big) +\\
&\quad c_{11}\left(\delta_{IL}\delta_{JP}\delta_{KQ} + \delta_{IP}\delta_{JL}\delta_{KQ}\right) +\\
&\quad c_{15}\left(\delta_{IL}\delta_{JQ}\delta_{KP} + \delta_{IP}\delta_{JQ}\delta_{KL} + \delta_{IQ}\delta_{JL}\delta_{KP} + \delta_{IQ}\delta_{JP}\delta_{KL}\right).
\end{aligned} \tag{1.3}$$

Small deformations in terms of the linear strain tensor $\boldsymbol{\epsilon} = \frac{1}{2}(\nabla\boldsymbol{u} + \boldsymbol{u}\nabla)$ are often sufficient to characterize the deformation behavior of solids. If the intention is to include deformation due to surface, line, or point stresses, gradients of this quantity are taken into account, for example, a gradient of displacement of the nth order, $\overset{(n)}{\boldsymbol{\epsilon}} =$

$\nabla^n \boldsymbol{u}$. For homogeneous, centrosymmetric, isotropic materials, the corresponding strain energy density is detailed in the seminal paper by Mindlin (1965) by taking into account all terms up to $\overset{(3)}{\epsilon} = \nabla\nabla\nabla\boldsymbol{u}$. Similar to the finite strain case, this comes at the expense of an enormous amount of stiffness coefficients besides the classical two Lamé constants, λ and μ. As indicated above, this makes it difficult for the experimentalist, because, just as in the case of Young's modulus and Poisson's ratio in Cauchy materials, independent experiments have to be conceived and then performed to obtain them. Moreover, very often some of them are unnecessary when the intention is to mimic certain phenomena that involve only a simple type of deformation or quasi-1D or quasi-2D continua are considered. It is therefore not surprising that the experimentally oriented literature of solids subjected to small deformations confines itself to a second-order theory and simplified variations thereof. For example in Lam et al. (2003), we find for the strain energy density of "strain gradient elasticity"

$$
\begin{aligned}
w = \tfrac{1}{2}\lambda\epsilon_{ii}\epsilon_{jj} + \mu\epsilon_{ij}\epsilon_{ij} + a_1 \overset{(2)}{\epsilon}_{ijj} \overset{(2)}{\epsilon}_{ikk} + a_2 \overset{(2)}{\epsilon}_{iik} \overset{(2)}{\epsilon}_{kjj} + \\
a_3 \overset{(2)}{\epsilon}_{iik} \overset{(2)}{\epsilon}_{jjk} + a_4 \overset{(2)}{\epsilon}_{ijk} \overset{(2)}{\epsilon}_{ijk} + a_5 \overset{(2)}{\epsilon}_{ijk} \overset{(2)}{\epsilon}_{kji},
\end{aligned}
\tag{1.4}
$$

where a_i denote the five additional second-order elastic constants associated with $\overset{(2)}{\epsilon} = \nabla\nabla\boldsymbol{u}$. Following Castigliano, it can be used to compute the hyperstresses

$$
\overset{(2)}{\boldsymbol{\sigma}} = \frac{\partial w}{\partial \epsilon}, \quad \overset{(3)}{\boldsymbol{\sigma}} = \frac{\partial w}{\partial \overset{(2)}{\epsilon}}. \tag{1.5}
$$

It is worth while mentioning that the Green–Lagrange strain tensor E_{IJ} can be linked to gradients of the displacement $\boldsymbol{u}$ and products thereof

$$
E_{IJ} = \tfrac{1}{2}\left(u_{I,J} + u_{J,I} + u_{K,I}u_{K,J}\right). \tag{1.6}
$$

Similarly, the gradient of $\boldsymbol{E}$ can be linked to second-order gradients of the displacement. Now if one restricts oneself to a linear theory in the sense that products of derivatives are neglected, Eqs. (1.2), (1.3), and (1.4) coincide.

Lam et al. (2003) also presented the following simplified version of the strain energy density, which restricts itself to three additional length scale parameters l_0, l_1, and l_2 sufficient for characterizing a nearly incompressible material

$$
w = \tfrac{1}{2}\lambda\epsilon_{ii}\epsilon_{jj} + \mu\epsilon_{ij}\epsilon_{ij} + \mu l_0^2 \epsilon_{mm,i}\epsilon_{nn,i} + \mu l_1^2 \overset{(2)}{\epsilon}{}'_{ijk} \overset{(2)}{\epsilon}{}'_{ijk} + \mu l_2^2 \overset{s}{\kappa}_{ij} \overset{s}{\kappa}_{ij} \tag{1.7}
$$

with the traceless symmetrized second gradient tensor

$$
\overset{(2)}{\epsilon}{}'_{ijk} = \overset{(2)}{\epsilon}{}^{s}_{ijk} - \tfrac{1}{5}\left(\delta_{ij} \overset{(2)}{\epsilon}{}^{s}_{mmk} + \delta_{jk} \overset{(2)}{\epsilon}{}^{s}_{mmi} + \delta_{ki} \overset{(2)}{\epsilon}{}^{s}_{mmj}\right), \tag{1.8}
$$

the symmetrized second gradient tensor

$$\overset{(2)}{\epsilon}{}^{\mathrm{s}}_{ijk} = \tfrac{1}{3}\left(\overset{(2)}{\epsilon}_{ijk} + \overset{(2)}{\epsilon}_{jki} + \overset{(2)}{\epsilon}_{kij}\right), \tag{1.9}$$

and the symmetrized wryness tensor

$$\kappa^{\mathrm{s}}_{ij} = \tfrac{1}{4}\left(\varepsilon_{ipq}\overset{(2)}{\epsilon}_{jpq} + \varepsilon_{jpq}\overset{(2)}{\epsilon}_{ipq}\right). \tag{1.10}$$

This approach is also referred to as "modified strain gradient theory."

The Experimental Facts

It is fair to say that the theoretical concepts outlined in the previous chapter led to a multitude of analytical and numerical solutions and that many initial-boundary value problems were solved. However, when it comes to compare these results with experiments or, even before that, when the question about numbers for the various new material constants arises, the situation is quite different. In what follows, we shall present some of the available experimental data.

Data and Results for Second Gradient Fluids

The groups of Trostel and Silber used the second gradient fluid theory presented in Trostel (1985), Silber et al. (2007, 1998) for studying the transition from laminar to turbulent flow of various water–oil mixtures. In their publications, absolute values for the constants ν_i in Eq. (1.1) are not revealed. Instead, the problem was studied in dimensionless form for Couette flow between two parallel plates at a distance h, where dimensionless material is defined as follows

$$\alpha_1 = \frac{\lambda_{vv}}{\kappa_1}\frac{h}{2}, \quad \alpha_2 = \frac{\lambda_{vv'}}{\kappa_1}, \quad \alpha_3 = \frac{\lambda_{v'v'}}{\kappa_1}\frac{2}{h}, \quad \Gamma = \lambda\frac{h}{2}, \quad \lambda^2 = \frac{\kappa_1}{\kappa_2}, \tag{1.11}$$

h being the height of the Couette flow channel and λ_{vv}, $\lambda_{vv'}$, $\lambda_{v'v'}$ are porosity coefficients arising from the boundary conditions. Corresponding data are shown in Table 1.2.

In Silber et al. (1998), the Poiseuille flow of blood and of suspensions of rigid spheres through (narrow) tubes was also studied and compared with external experimental results. A relation for the stationary flow profile was derived

$$v(\eta) = v_0\left[1 - \eta^2 + 2\frac{1+\xi}{\Lambda^2}\frac{I_0(\Lambda\eta) - I_0(\Lambda)}{I_0(\Lambda) - \left(1 - \frac{\kappa_3}{\kappa_2}\right)\frac{I_1(\Lambda)}{\Lambda}}\right], \tag{1.12}$$

Table 1.1 Second gradient material parameters for blood and water/rigid spheres

Experimental Ref.	$Q\,(\mathrm{mm}^3/\mathrm{s})$	B (mm/s)	Λ	ξ
Goldsmith and Turitto (1986)	1.96×10^{-3}	0.03886	−4.93	−4.356
Karnis et al. (1966)	7.11	N/A	2.236	−3.543
Karnis et al. (1966)	35.6	N/A	2.24	−3.338

Table 1.2 Second gradient material parameters for water/water–oil

Reference	Experimental Ref.	α_1	α_2	α_3	Γ
Silber et al. (1998)	Reichardt (1956)	0.011	0.094	0.866	5.26
Silber et al. (2007)	Piesche (1983)	5.0861629	1.9319036	0.73508637	6.944230
Alizadeh et al. (2011)	Hagiwara et al. (2002)	250	8.4038	0.28334	5.75755

where

$$\eta = \frac{r}{R}, \quad v_0 = \frac{p'}{\kappa_1}\left(\frac{R}{2}\right)^2, \quad \xi = \frac{\kappa_3 - \lambda_{v'v'} R}{\kappa_2}, \quad \Lambda = \lambda R. \tag{1.13}$$

r is the radial coordinate within the circular tube, R is its external radius, p' is the pressure gradient, and I_0 and I_1 denote Bessel functions of the first kind of order zero and one. All relevant parameters are compiled in Table 1.1. The reader of the original paper by Silber et al. (1998) should be aware that it contains typographical errors, which were corrected in the present article. Moreover, data for the ratio κ_3/κ_2 are not mentioned. The plots shown in that reference were therefore recalculated. After that, it seemed likely that the porosity coefficient $\lambda_{v'v'}$ was chosen to be zero. Hence, $\xi = \kappa_3/\kappa_2$ (Table 1.2).

Figure 1.1a shows the flow profile of blood predicted by means of second gradient theory (blue curve). It agrees well with the experimental data in contrast to the traditional parabolic profile (red curve).

Figure 1.1b shows the flow profile of glass spheres in water predicted by means of second gradient theory. The blue and red curves relate to small and large values of flow rate Q, respectively. Both agree well with experimental data.

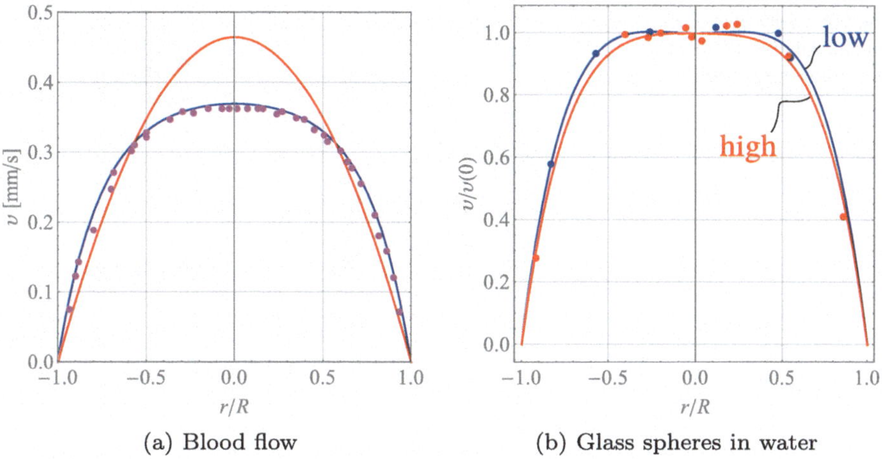

Fig. 1.1 Velocity profiles of Poiseuille flow of blood and of suspensions of rigid spheres through a tube. Comparison of models and experimental data from Silber et al. (1998)

Data and Results for Second Gradient Solids

In his work, Chong (2002) considers three different types of application of strain gradient theory: (a) nanoindentation, (b) bending of wide thin plates, and (c) cantilever bending. He concentrates on micro-scaled experiments with a cured commercially available liquid epoxy resin, namely Shell Epikote 828 epoxy resin, bisphenol A epichlorohydrin. In his experiments, an (elastic) size effect can be observed: The stiffness of the material seems to increase as indentation depth or specimen size decreases. However, the physics behind this observation is not really revealed. We only find tautological remarks, such as "... The increase in hardness was attributed to the increasing contribution of the strain gradients to the hardness when indent size decreases. ..." Results from nanoindentation tests using different indentation pyramids are shown in Fig. 1.2. Note that the "stiffness" or (more accurately) the reduced Young's modulus can be obtained from the slope of the unloading path during an indentation test.

In the plot, the slope arising at small indentation depths was normalized w.r.t. the slope encountered during deep depth probing. The size effect becomes visible at depths smaller than $1\,\mu$m. Below that, the microstructure of the epoxy must take effect.

Only parametric studies and no experimental data are reported for the case (b) of plate bending. However, case (c) for the bending of a simply supported cantilever was studied experimentally after fabricating micron-sized beams of thicknesses 20, 38, 75, and 115 microns. The "macroscopic" Young's modulus and Poisson's ratio were determined from tensile tests, where no size effect was observed, as it should be because there are no strain gradients. They were used in the following equation

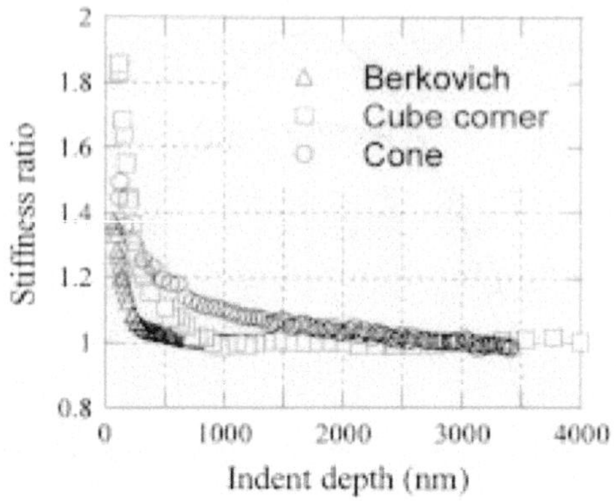

Fig. 1.2 Size effect during nanoindentation of epoxy; plot from Chong (2002)

for the maximum deflection, $w_{\max}$, for fitting the length scale parameters l_0, l_1, l_2 of the modified strain gradient approach, which in addition were assumed to be equal

$$w_{\max} = \frac{1}{3}\frac{Fl^3}{Dbh^3}. \tag{1.14}$$

l is the span, h the height, b the width, and F the force applied to the beam. Moreover, the apparent bending stiffness, D', is given by

$$D' = D_0\frac{1}{3}\left[1 + \left(\frac{b_h}{h}\right)^2\right], \quad D_0 = \frac{E}{12\alpha}, \tag{1.15}$$

where $b_h^2 = 6(1-2\nu)l_0^2 + \frac{2}{5}(4-\nu)l_1^2 + 3(1-\nu)l_2^2$. Moreover, $\alpha = 1-\nu^2$ and $\alpha = 1$ for plane stress and plane strain, respectively.

Some results are shown in Fig. 1.3. Clearly, the size effect is visible only if the intrinsic length scale, b_h, and the height of beam, h (along which strain gradients are present), are comparable. This seems to be the case below 80 microns.

Li et al. (2009) presented similar results for carbon nanotubes (CNTs). Figure 1.4 shows the variation in effective Young's modulus with diameter for different internal lengths. Here, the size effect sets in at an even smaller scale, which is obviously related to the size of the tested structure.

In Liebold (2015), Liebold and Müller (2015), and Liebold and Müller (2016), size effect results from external and own experiments were compiled in tabular form, which for convenience are repeated here: Tables 1.3 and 1.4.

Lei et al. (2016) used dynamic test methods and studied the eigenvibrations of micrometer-size nickel beams shown in Fig. 1.5.

Some of their experimental results and fitting curves based on strain gradient theory (SGT) and modified couple stress theory (MCST) are shown in Fig. 1.6. The stiffening effect becomes obvious from the increase of the first eigenfrequency of the simply supported beam. However, a microscopic explanation of this effect was

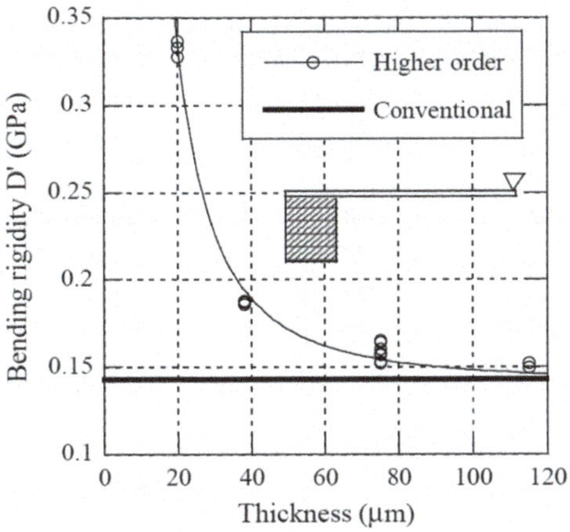

Fig. 1.3 Size effect when bending epoxy beams; plot from Chong (2002) and Lam et al. (2003)

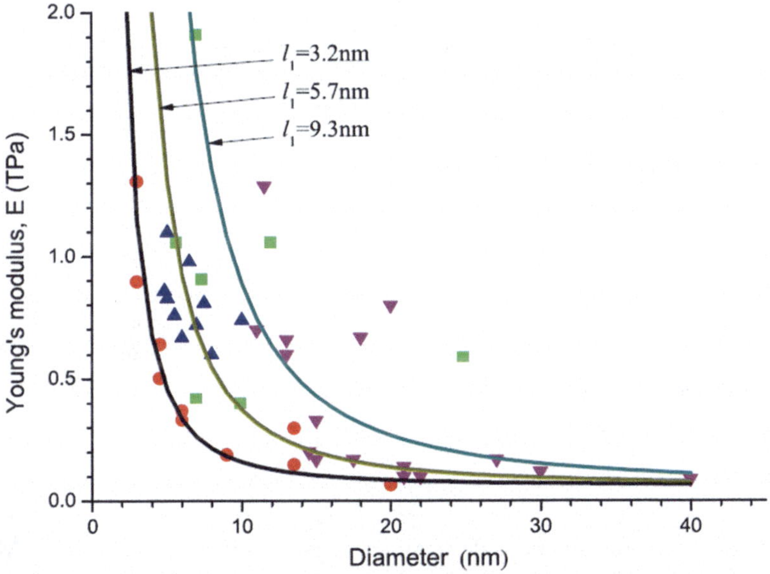

Fig. 1.4 Size effect when bending CNTs; comparison of prediction experimental data for Young's modulus as a function of the diameter; plot from Li et al. (2009)

Table 1.3 Some literature values for the size effect in elasticity from Liebold and Müller (2016). SGT: Strain gradient theory; MCST: modified couple stress theory

Length scale l	Material	Theory	Testing method	Reference
$l = 9.4\,\mu m$	Epoxy	CS	Bending bars	Chong (2002), Lam et al. (2003)
$l = 3.0\,\mu m$	Copper	MSG	Torsion wires	Yang et al. (2002)
$l = 7.0\,nm$	Zinc oxide	MSG	Torsion	Stan et al. (2007)
$l = 57.0\,nm$	Polypyrrole nanotubes	MSG	Bending	Cuenot et al. (2000)
$l = 3.2\,nm$	Single-walled CNT	SG	Bending	Jing et al. (2006), Sadeghian et al. (2009)
$l = 5.7\,nm$	Multi-walled CNT	SG	Bending	Jing et al. (2006), Sadeghian et al. (2009)
$l = 9.3\,nm$	Multi-walled CNT	SG	Bending	Poncharal et al. (1999), Jing et al. (2006)
$l = 17.7\,nm$	Lead	CS	Bending	Cuenot et al. (2004)

Table 1.4 Size effect in elasticity (own measurements)

Material	Theory	Young's modulus E (GPa)	Length scale l (μm)
Epoxy	CS	3.93	7.75
	MSG	3.93	4.35
SU-8	CS	4.13	2.5
	MSG	4.14	1.39
Drilled aluminum	CS	39.8	920
	MSG	39.8	510

not provided. Several length scale parameters are reported in their work. May it suffice to say that the main conclusion was that for the SGT microbeam model, the dilatation gradient has the greatest influence on the microbeams followed by the rotation gradient and finally the deviatoric stretch gradient. For the numerical parameters, the reader is referred to the paper.

Figure 1.8 (left inset) shows the fit based on couple stress and modified strain gradient theory. The corresponding length scale parameters are already mentioned in Table 1.4. Another fit based on the so-called surface elasticity/core–shell model is also shown. For conciseness, this approach was not explained in the present work. However, what should be mentioned is the fact that this theory may serve to model the negative size effect observed in the foams.

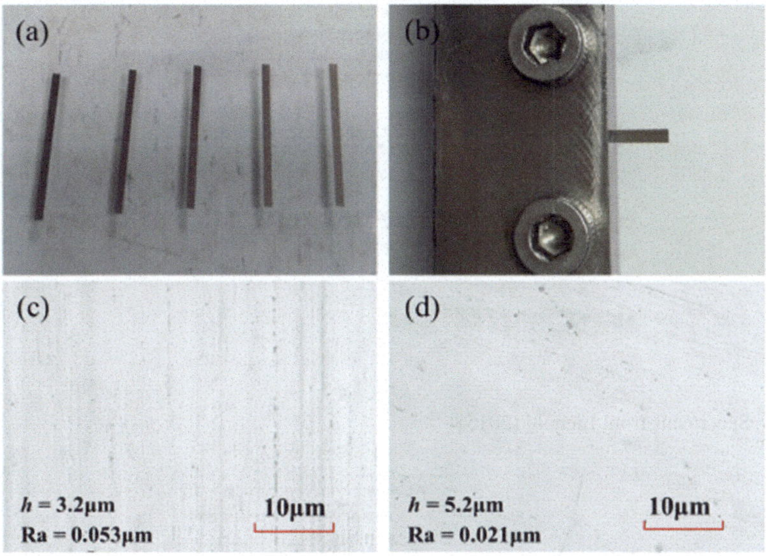

Fig. 1.5 Specimens from Lei et al. (2016): **a** nickel stripes; **b** nickel cantilever microbeam; **c** and **d** surface morphology and the roughness of the foils

Fig. 1.6 Data from Lei et al. (2016): mean values of the experimental results, SGT solution and MCST solution. Dotted and solid line coincide

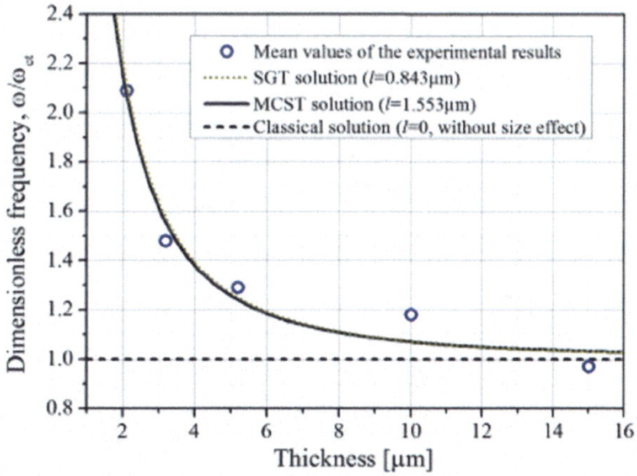

It should be noted that similar dynamic measurements (for double suspended beams) have been presented in Liebold (2015) and Liebold and Müller (2016) to study size effects in aluminum with drilled holes and foams: Fig. 1.7 (Fig. 1.8).

The strain gradient theory was applied in Réthoré et al. (2015) to study the deformation of a 3D-printed honeycomb structure containing an inclined crack: Fig. 1.9.

No detailed information about the printer material is provided. A micro-sectional analysis showing that the printed conglomerate can truly be considered to be an elastic continuum is also missing. Nevertheless, the authors used digital image correlation

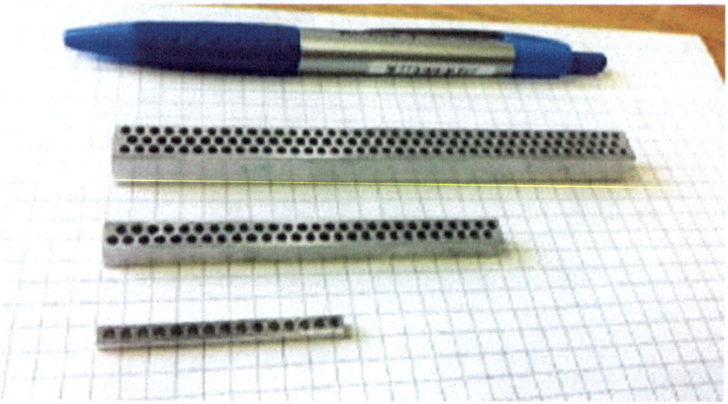

Fig. 1.7 Specimens from Liebold (2015)

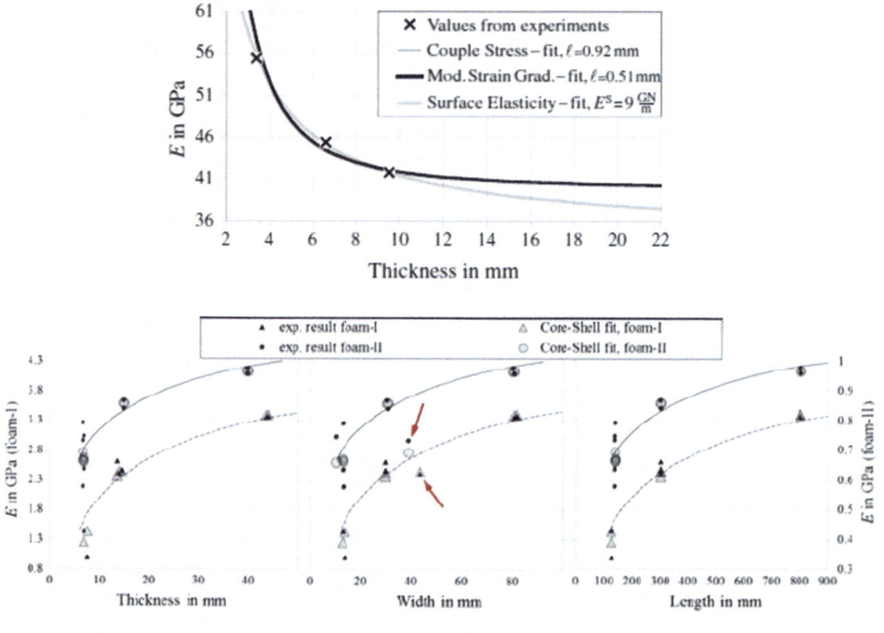

Fig. 1.8 Size effect in drilled aluminum bars and aluminum foams from Liebold and Müller (2016)

in combination with a finite element analysis to determine the additional five elastic constants of second gradient elasticity. The procedure is illustrated in Fig. 1.10.

The school of Dell'Isola, e.g., in Turco et al. (2016) developed the concept of a so-called pantographic structure, which is the epitome of a realization of a second-order gradient material documented by mathematical proof (see Dell'Isola et al. 2019 or Barchiesi et al. 2018): Fig. 1.11.

Such structures are realized by means of 3D-printing and were so far made of polyamide, aluminum, and titanium. Their strength is a huge propensity for huge

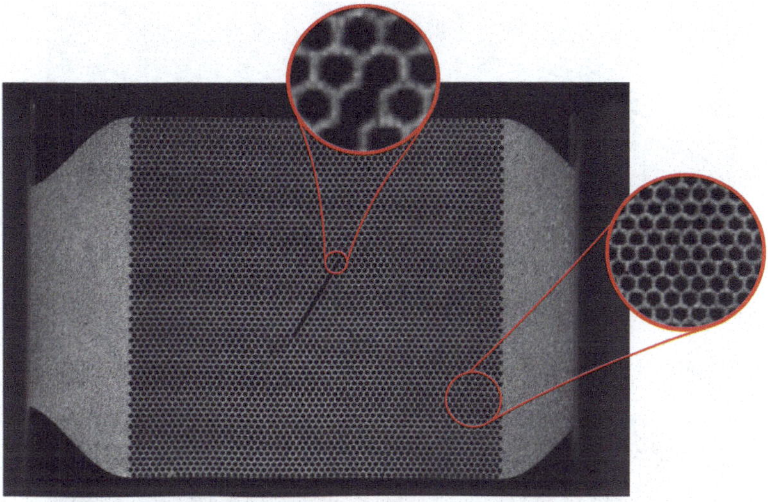

Fig. 1.9 3D-printed specimen containing an inclined crack; adapted from Réthoré et al. (2015)

Fig. 1.10 Digital image correlation and related FE simulation from Réthoré et al. (2015): horizontal displacement field in pixel obtained with the mesh used for the identification of the material constants

damage tolerance even at large strains, Fig. 1.12, which makes them strong candidates as buffer materials for crash absorption.

As a first step toward the determination of the extended set of material parameters, digital image correlation was applied in order to determine the displacements fully in 3D: Fig. 1.13.

In a next step, the DIC data are correlated with FE simulations by inverse analysis to obtain the best parameter fit. How to define the best experiment for the various parameters was examined in Placidi et al. (2015) based on the theory for second

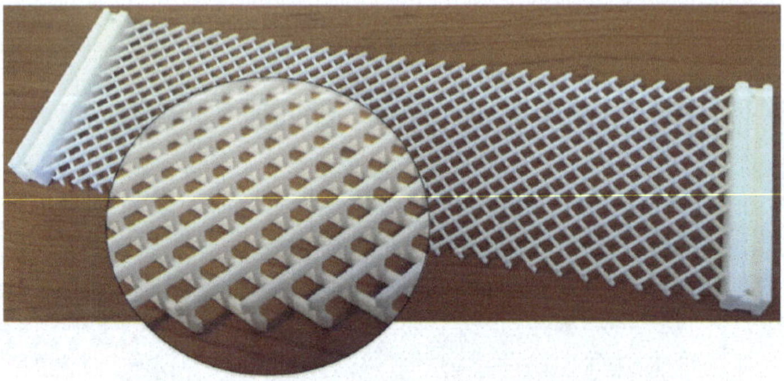

Fig. 1.11 3D-printed polyamide pantographs from Dell'Isola et al. (2019)

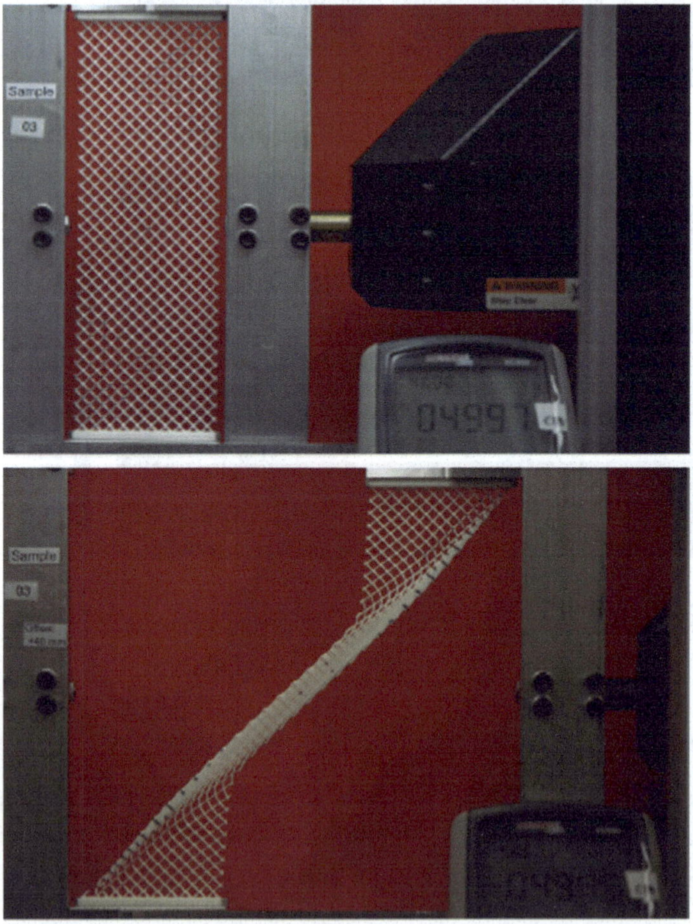

Fig. 1.12 Shear deformation of pantographs from Turco et al. (2016)

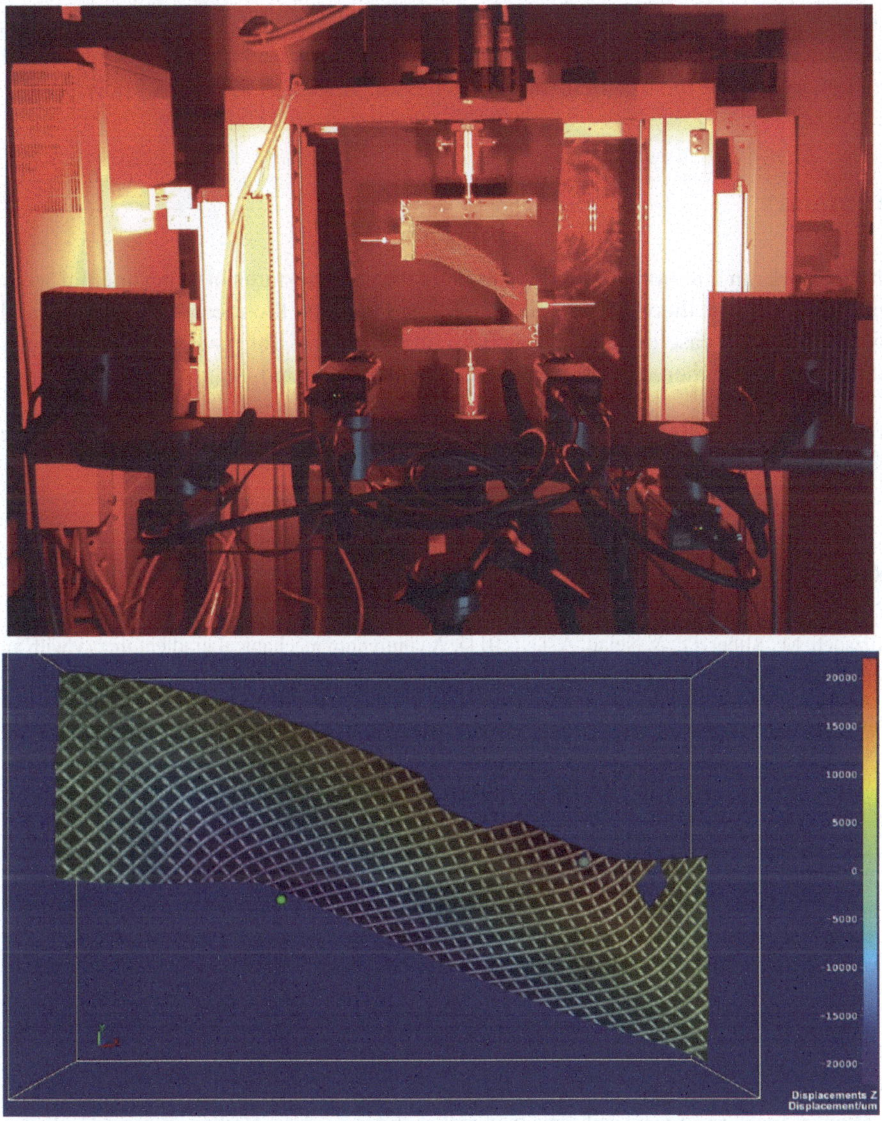

Fig. 1.13 DIC performed on pantographs from Ganzosch et al. (2018)

gradient materials presented above. However, in Yang and Müller (2019) it was shown that second gradient elasticity with small deformation measures might not be enough to capture the true deformation behavior, at least in the case of polyamide.

Summary

In this work, an attempt was made to present the current status of experimental work on higher gradient theories. Specifically, second gradient theories were reviewed and explored for the case of fluids as well as solids. Results for material data from the literature were gathered, presented, and critically evaluated.

Acknowledgements The author wants to thank Mr. Wilhelm Rickert, M.Sc., for his most valuable assistance in the production of this contribution.

References

Alizadeh, M., Silber, G., & Nejad, A. G. (2011). A continuum mechanical gradient theory with an application to fully developed turbulent flows. *Journal of Dispersion Science and Technology*, *32*(2), 185–192.

Barchiesi, E., Dell'Isola, F., Laudato, M., Placidi, L., & Seppecher, P. (2018). A 1D continuum model for beams with pantographic microstructure: Asymptotic micro-macro identification and numerical results. In F. DellŹIsola, V. A. Eremeyev, & A. Porubov, (Eds.), *Advances in mechanics of microstructured media and structures* (pp. 43–74). Cham: Springer.

Bertram, A. (2019). *Compendium on gradient materials including solids and fluids*. Internet, Magdeburg, Berlin. https://www.lkm.tu-berlin.de/fileadmin/fg49/publikationen/bertram/Compendium_on_Gradient_Materials_June_2019.pdf.

Chong, C. M. (2002). *Experimental investigation and modeling of size effect in elasticity*. Ph.D. thesis, University of Science and Technology, Hong Kong.

Cuenot, S., Demoustier-Champagne, S., & Nysten, B. (2000). Elastic modulus of polypyrrole nanotubes. *Physical Review Letters*, *85*(8), 1690–1693.

Cuenot, S., Frétigny, C., Demoustier-Champagne, S., & Nysten, B. (2004). Surface tension effect on the mechanical properties of nanomaterials measured by atomic force microscopy. *Physical Review B*, *69*(16), 165410.

Dell'Isola, F., Corte, A. D., & Giorgio, I. (2017). Higher-gradient continua: The legacy of Piola, Mindlin, Sedov and Toupin and some future research perspectives. *Mathematics and Mechanics of Solids*, *22*(4), 852–872.

Dell'Isola, F., Sciarra, G., & Vidoli, S. (2009). Generalized Hooke's law for isotropic second gradient materials. *Proceedings of the Royal Society A: Mathematical, Physical and Engineering Sciences*, *465*(2107), 2177–2196.

Dell'Isola, F., Seppecher, P., Alibert, J. J., Lekszycki, T., Grygoruk, R., & Pawlikowski, M., et al. (2019). Pantographic metamaterials: An example of mathematically driven design and of its technological challenges. *Continuum Mechanics and Thermodynamics*, *31*(4), 851–884.

Dell'Isola, F., Seppecher, P., & Corte, A. D. (2015). The postulations á la D'Alembert and á la Cauchy for higher gradient continuum theories are equivalent: a review of existing results. *Proceedings of the Royal Society A: Mathematical, Physical and Engineering Sciences*, *471*(2183), 20150415.

Ganzosch, G., Hoschke, K., Lekszycki, T., Giorgio, I., Turco, E., & Müller, W. (2018). 3D-measurements of 3D-deformations of pantographic structures. *Technische Mechanik, 38*(3), 233–245.

Goldsmith, H. L., & Turitto, V. T. (1986). Rheological aspects of thrombosis and haemostasis: Basic principles and applications. *Thrombosis and Haemostasis, 56*(3), 415–435.

Hagiwara, Y., Sakamoto, S., Tanaka, M., & Yoshimura, K. I. (2002). Ptv measurement on interaction between two immiscible droplets and turbulent uniform shear flow of carrier fluid. *Experimental Thermal and Fluid Science, 26*(2–4), 245–252.

Jing, G. Y., Duan, H., Sun, X. M., Zhang, Z. S., Xu, J., & Li, Y. D., et al. (2006). Surface effects on elastic properties of silver nanowires: Contact atomic-force microscopy. *Physical review B, 73*(23), 235409.

Karnis, A., Goldsmith, H. L., Mason, S. G. (1966). The kinetics of flowing dispersions: I. Concentrated suspensions of rigid particles. *Journal of Colloid and Interface Science, 22*(6), 531–553.

Lam, D. C., Yang, F., Chong, A. C. M., Wang, J., & Tong, P. (2003). Experiments and theory in strain gradient elasticity. *Journal of the Mechanics and Physics of Solids, 51*(8), 1477–1508.

Lei, J., He, Y., Guo, S., Li, Z., & Liu, D. (2016). Size-dependent vibration of nickel cantilever microbeams: Experiment and gradient elasticity. *AIP Advances, 6*(10), 105202 1–7.

Li, X.-F., Wang, B.-L., & Kang, Y. L. (2009). Size effects of the bending stiffness of nanowires. *Journal of Applied Physics, 105*(7), 074306–1–6.

Liebold, C. (2015). *Größeneffekt in der Elastizität*. Ph.D. thesis, Technische Universität Berlin, Berlin.

Liebold, C., & Müller, W. H. (2015). Applications of strain gradient theories to the size effect in submicro-structures incl. experimental analysis of elastic material parameters. *Bulletin of TICMI, 19*(1), 45–55.

Liebold, C., & Müller, W. H. (2016). Comparison of gradient elasticity models for the bending of micromaterials. *Computational Materials Science, 116*, 52–61.

Mindlin, R. D. (1965). Second gradient of strain and surface-tension in linear elasticity. *International Journal of Solids and Structures, 1*(4), 417–438.

Piesche, M. (1983). Strömungs- und Wärmetransportvorgänge im Einlaufbereich eines ebenen Spaltes mit relativ zueinander bewegten Wänden. *Strömungsmechanik und Strömungsmaschinen, 33*, 1–58.

Placidi, L., Andreaus, U., Della Corte, A., & Lekszycki, T. (2015). Gedankenexperiments for the determination of two-dimensional linear second gradient elasticity coefficients. *Zeitschrift für angewandte Mathematik und Physik, 66*(6), 3699–3725.

Poncharal, P., Wang, Z. L., Ugarte, D., & De Heer, W. A. (1999). Electrostatic deflections and electromechanical resonances of carbon nanotubes. *Science, 283*(5407), 1513–1516.

Reichardt, H. (1956). Über die Geschwindigkeitsverteilung in einer geradlinigen turbulenten Couettieströmung. *Zeitschrift für Angewandte Mathematik und Mechanik, 36*(S1), S26–S29.

Réthoré, J., Kaltenbrunner, C., Dang, T. B. T., Chaudet, P., & Kuhn, M. (2015). Gradient-elasticity for honeycomb materials: Validation and identification from full-field measurements. *International Journal of Solids and Structures, 72*, 108–117.

Sadeghian, H., Yang, C. K., Goosen, J. F. L., Van Der Drift, E., Bossche, A., & French, P. J., et al. (2009). Characterizing size-dependent effective elastic modulus of silicon nanocantilevers using electrostatic pull-in instability. *Applied Physics Letters, 94*(22), 235409.

Silber, G., Janoske, U., Alizadeh, M., & Benderoth, G. (2007). An application of a gradient theory with dissipative boundary conditions to fully developed turbulent flows. *Journal of Fluids Engineering, 129*(5), 643–651.

Silber, G., Trostel, R., Alizadeh, M., & Benderoth, G. (1998). A continuum mechanical gradient theory with applications to fluid mechanics. *Le Journal de Physique IV, 8*(PR8), Pr8–365–Pr8–373.

Stan, G., Ciobanu, C. V., Parthangal, P. M., & Cook, R. F. (2007). Diameter-dependent radial and tangential elastic moduli of zno nanowires. *Nano Letters, 7*(12), 3691–3697.

Trostel, R. (1985). Gedanken zur Konstruktion mechanischer Theorien. Beiträge zu den Ingenieurwissenschaften, pp. 96–134. Univ.-Bibl. Techn. Univ. Berlin.
Turco, E., Dell'Isola, F., Rizzi, N. L., Grygoruk, R., Müller, W. H., & Liebold, C. (2016). Fiber rupture in sheared planar pantographic sheets: Numerical and experimental evidence. *Mechanics Research Communications*, *76*, 86–90.
Yang, F. A. C. M., Chong, A. C. M., Lam, D. C. C., & Tong, P. (2002). Couple stress based strain gradient theory for elasticity. *International Journal of Solids and Structures*, *39*(10), 2731–2743.
Yang, H., & Müller, W. H. (2019). Computation and experimental comparison of the deformation behavior of pantographic structures with different micro-geometry under shear and torsion. *Journal of Theoretical and Applied Mechanics*, *57*(2), 421–434.

Chapter 2
Balance Laws for Gradient Materials

Albrecht Bertram

Abstract If we extend the kinematical variables to higher gradients, then we also have to extend the dynamical ones to higher-order stress fields. For them, an extension of the balances and the boundary conditions is required. There are different ways to achieve this. Here, we suggest one based on an objectivity assumption of the power. The power functional is extended to all virtual velocity fields in a linear and continuous way. By applying the Riesz representation to the virtual power, we can introduce hyperstresses. The exploitation of the Euclidean transformation leads to forces and torques. Integral transformations give us finally the form of the balance laws including the boundary terms determining the possible boundary conditions (For more details, see Bertram: *Compendium on Gradient Materials* (various editions)).

Tensor Notations

We denote the order of the tensors like $\overset{\langle k \rangle}{C}$ for a kth-order tensor. For a second-order tensor or a dyadic, we eventually write simply $\mathbf{C}$ for $\overset{\langle 2 \rangle}{C}$.

For every contraction between tensors, we put one dot. More exactly, the ***P*-fold contraction** of a K-fold tensor product $\mathbf{v}_1 \otimes \ldots \otimes \mathbf{v}_K$ with an M-fold tensor product $\mathbf{x}_1 \otimes \ldots \otimes \mathbf{x}_M$ for $K \geq P \leq M$ is the ($K + M$–$2P$)-fold tensor product

$$(\mathbf{v}_1 \otimes \ldots \otimes \mathbf{v}_K) \cdot \ldots \cdot (\mathbf{x}_1 \otimes \ldots \otimes \mathbf{x}_M)$$
$$= \mathbf{v}_1 \otimes \ldots \otimes \mathbf{v}_{K-P} \otimes \mathbf{x}_{P+1} \otimes \ldots \otimes \mathbf{x}_M (\mathbf{v}_{K-P+1} \cdot \mathbf{x}_1)(\mathbf{v}_{K-P+2} \cdot \mathbf{x}_2) \ldots (\mathbf{v}_K \cdot \mathbf{x}_P). \tag{2.1}$$

Otto-von-Guericke-University Magdeburg

A. Bertram (✉)
Technical University Berlin and Otto-von-Guericke-University, Magdeburg, Germany
e-mail: albrecht.bertram@ovgu.de

A. Bertram and S. Forest (eds.), *Mechanics of Strain Gradient Materials*,
CISM International Centre for Mechanical Sciences 600,
https://doi.org/10.1007/978-3-030-43830-2_2

wherein "$\cdot\ldots\cdot$" stands for P contraction dots. For better visibility, one can eventually arrange these contraction dots in groups with identical meaning, like $\therefore$ for $\cdots$, and :: for $\cdots\cdot$.

A triadic is **right subsymmetric** if for the components with respect to an orthonormal vector basis $A_{ijk} = A_{ikj}$, and **left subsymmetry** if $A_{ijk} = A_{jik}$.

The **Rayleigh product** maps all basis vectors of a tensor simultaneously without changing its components. To be more precise, let $\overset{\langle k\rangle}{\boldsymbol{C}}$ be a tensor of kth order ($k \geq 1$) and $\mathbf{T}$ a dyadic. Then, the Rayleigh product between them is defined as

$$\begin{aligned}\mathbf{T} * \overset{\langle k\rangle}{\boldsymbol{C}} &= \mathbf{T} * (C^{i_1 i_2 \cdots i_k} \mathbf{r}_{i_1} \otimes \mathbf{r}_{i_2} \otimes \ldots \otimes \mathbf{r}_{i_k}) \\ &:= C^{i_1 i_2 \cdots i_k} (\mathbf{T} \cdot \mathbf{r}_{i_1}) \otimes (\mathbf{T} \cdot \mathbf{r}_{i_2}) \otimes \ldots \otimes (\mathbf{T} \cdot \mathbf{r}_{i_k}).\end{aligned} \tag{2.2}$$

Kinematics

In this section, we introduce the basic kinematical concepts of continuum mechanics.

A **motion** of a body in the Euclidean space is described by a function

$$\mathbf{x} = \boldsymbol{\chi}(\mathbf{x}_0, t) \tag{2.3}$$

which assigns at each instant of time t the position vector of a particle identified by its position vector $\mathbf{x}_0$ in the reference placement to its position $\mathbf{x}$ in the current placement. This function is assumed to be continuous and differentiable in space and time.

The volume and surface elements in the current placement are $\mathrm{d}V$ and $\mathrm{d}A$, and in the reference placement $\mathrm{d}V_0$ and $\mathrm{d}A_0$, respectively. For the mass element $\mathrm{d}m$, a distinction of the two placements is not necessary because of mass conservation. The mass densities in the two placements are denoted by ρ and ρ_0, respectively.

The **displacement** field is defined as the difference vector of the position vectors in the current and reference placement describing the same particle

$$\mathbf{u}(\mathbf{x}_0, t) = \boldsymbol{\chi}(\mathbf{x}_0, t) - \mathbf{x}_0. \tag{2.4}$$

The partial differential with respect to time (denoted by a dot) is the **velocity** field

$$\mathbf{v}(\mathbf{x}_0, t) = \frac{\partial \boldsymbol{\chi}}{\partial t} = \boldsymbol{\chi}(\mathbf{x}_0, t)^{\bullet} \tag{2.5}$$

and the second time derivative is the **acceleration** field

$$\mathbf{a}(\mathbf{x}_0, t) = \boldsymbol{\chi}(\mathbf{x}_0, t)^{\bullet\bullet}. \tag{2.6}$$

When using spatial derivatives or gradients of such fields, we have to take care of two different choices. So grad, div and curl are related to the gradient, divergence and curl operation, respectively, in the current placement, while Grad, Div and Curl are related to the same operations in the reference placement.

The material derivative with respect to the position vector is called **deformation gradient**

$$\mathbf{F}(\mathbf{x}_0, t) = \frac{\partial \boldsymbol{\chi}}{\partial \mathbf{x}_0} = \operatorname{Grad} \boldsymbol{\chi}(\mathbf{x}_0, t) = \boldsymbol{\chi}(\mathbf{x}_0, t) \otimes \nabla_0 \tag{2.7}$$

giving an invertible tensor field.

In contrast to the material nabla ∇_0 acting in the reference placement, the spatial nabla acting in the current placement is denoted by ∇. After the chain rule, the two nablas are transformed as

$$\nabla_0 = \nabla \cdot \mathbf{F} = \mathbf{F}^T \cdot \nabla \quad \text{and} \quad \nabla = \nabla_0 \cdot \mathbf{F}^{-1} = \mathbf{F}^{-T} \cdot \nabla_0. \tag{2.8}$$

The **displacement gradient** is

$$\mathbf{H} := \operatorname{Grad} \mathbf{u}(\mathbf{x}_0, t) = \mathbf{F} - \mathbf{I} \tag{2.9}$$

The linear **strain tensor** is its symmetric part

$$\mathbf{E} := 1/2\big(\mathbf{H} + \mathbf{H}^T\big) = 1/2\big(\mathbf{F} + \mathbf{F}^T - 2\mathbf{I}\big) \tag{2.10}$$

which plays an important role in the small deformation theory.

The (spatial) **velocity gradient** is

$$\mathbf{L} = \operatorname{grad} \mathbf{v} = \mathbf{v} \otimes \nabla = \operatorname{grad} \boldsymbol{\chi}^{\bullet} \tag{2.11}$$

which is in general non-symmetric. It can be decomposed into its symmetric part $\mathbf{D}$ and its skew part $\mathbf{W}$ as

$$\mathbf{L} = \mathbf{D} + \mathbf{W}. \tag{2.12}$$

$\mathbf{D}$ is called rate of deformation tensor and $\mathbf{W}$ spin tensor.

The time derivative of the deformation gradient is related to the velocity gradient by

$$\mathbf{L} = \mathbf{F}^{\bullet} \cdot \mathbf{F}^{-1}. \tag{2.13}$$

In the theory of finite deformations, the following symmetric tensors are often used. The **right Cauchy–Green tensor** is

$$\mathbf{C} := \mathbf{F}^T \cdot \mathbf{F} \tag{2.14}$$

and the **left Cauchy–Green tensor**

$$\mathbf{B} := \mathbf{F} \cdot \mathbf{F}^T \tag{2.15}$$

and **Green's strain tensor**

$$\mathbf{E}^{\mathrm{G}} := 1/2(\mathbf{C} - \mathbf{I}) \tag{2.16}$$

such that

$$\mathbf{E}^{\mathrm{G}\bullet} = 1/2\,\mathbf{C}^\bullet = \mathbf{F}^T * \mathbf{D}. \tag{2.17}$$

Observer dependence. Almost all of these kinematical concepts depend on the particular observer (or frame of reference). Even if all of them agree to use a joint reference placement, every observer will place the position vector to the current placement of some material point P in a different way. This is expressed by the **Euclidean transformation** between such position vectors $\mathbf{x}$ and $\mathbf{x}^*$

$$\mathbf{x}^*(P,t) = \mathbf{Q}(t) \cdot \mathbf{x}(P,t) + \mathbf{c}(t) \tag{2.18}$$

by a time-dependent vector $\mathbf{c}(t)$ and a time-dependent orthogonal tensor $\mathbf{Q}(t)$, both of which are determined solely by the two observers, but are independent of the motion of P. This transformation is the general isometry in the Euclidean space which means that the distance between two points P_1 and P_2 for all times is independent of the observer

$$\left|\mathbf{x}^*(P_1,t) - \mathbf{x}^*(P_2,t)\right| = |\mathbf{x}(P_1,t) - \mathbf{x}(P_2,t)| \tag{2.19}$$

which can be easily verified by (2.18).

If we use the Euclidean transformation to determine the velocity of a particle with respect to two observers, we will obtain the relation

$$\begin{aligned} \mathbf{v}^* &= \mathbf{Q} \cdot \mathbf{v} + \mathbf{Q}^\bullet \cdot \mathbf{Q}^T \cdot \left(\mathbf{x}^* - \mathbf{c}\right) + \mathbf{c}^\bullet \\ &= \mathbf{Q} \cdot \mathbf{v} + \boldsymbol{\omega} \times \left(\mathbf{x}^* - \mathbf{c}\right) + \mathbf{c}^\bullet \end{aligned} \tag{2.20}$$

where the angular velocity $\boldsymbol{\omega}(t)$ is the axial vector of the skew tensor $\mathbf{Q}(t)^\bullet \cdot \mathbf{Q}(t)^T$

$$\boldsymbol{\omega} := axi\left(\mathbf{Q}^\bullet \cdot \mathbf{Q}^T\right). \tag{2.21}$$

The transformation of the acceleration becomes even more complicated

$$\mathbf{a}^* = \mathbf{Q} \cdot \mathbf{a} + \mathbf{c}^{\bullet\bullet} + \boldsymbol{\omega}^\bullet \times \left(\mathbf{x}^* - \mathbf{c}\right) + \boldsymbol{\omega} \times \left[\left(\mathbf{x}^* - \mathbf{c}\right) \times \boldsymbol{\omega}\right] + 2\boldsymbol{\omega} \times \left(\mathbf{v}^* - \mathbf{c}^\bullet\right) \tag{2.22}$$

as a sum of the relative acceleration, the translational acceleration, the angular acceleration, the centripetal acceleration and the Coriolis acceleration.

The velocity gradient transforms as

$$\mathbf{L}^* = \mathbf{Q} * \mathbf{L} + \mathbf{Q}^\bullet \cdot \mathbf{Q}^T \tag{2.23}$$

and its skew part as

$$\mathbf{W}^* = \mathbf{Q} * \mathbf{W} + \mathbf{Q}^\bullet \cdot \mathbf{Q}^T \tag{2.24}$$

while the symmetric part is transformed as

$$\mathbf{D}^* = \mathbf{Q} * \mathbf{D} = \mathbf{Q} \cdot \mathbf{D} \cdot \mathbf{Q}^T. \tag{2.25}$$

We call tensor quantities **T** with such a transformation behavior **objective**

$$\mathbf{T}^* = \mathbf{Q} * \mathbf{T} \tag{2.26}$$

So, **D** is an objective tensor, and so are all higher velocity gradients from the second one upward. This definition can also be analogously applied to vectors (as first-order tensors) and scalars (as zeroth-order tensors). For scalars, the notions of objectivity and invariance coincide.

In this sense, the deformation gradient transforms neither as an objective nor as an invariant tensor since

$$\mathbf{F}^* = \mathbf{Q} \cdot \mathbf{F}. \tag{2.27}$$

The right Cauchy–Green tensor turns out to be invariant

$$\mathbf{C}^* = \mathbf{C} \tag{2.28}$$

and so is Green's strain tensor

$$\mathbf{E}^{\mathrm{G}*} = \mathbf{E}^{\mathrm{G}} \tag{2.29}$$

while the left Cauchy–Green tensor is objective

$$\mathbf{B} = \mathbf{Q} * \mathbf{B}. \tag{2.30}$$

Dynamics

Up to now, we have only considered kinematical quantities. All of them are visible quantities since they can be measured from a 3D video of the motion, at least in principle, by using a yardstick, a goniometer and a stopwatch. In contrast, all dynamical quantities to follow in our analysis will be invisible and thus not (directly) measurable, although they are the reasons for particular motions of bodies after our deterministic philosophical ansatz.

Classical approach after Euler and Cauchy. The starting point of the great pioneers in (continuum) mechanics like Newton, Euler and Cauchy was the balance of linear momentum and of moment of momentum. The general form of them is

$$\mathbf{l}^{\bullet} = \mathbf{f} \tag{2.31}$$

and

$$\mathbf{d}_O{}^{\bullet} = \mathbf{m}_O \tag{2.32}$$

with the linear momentum

$$\mathbf{l} := \int_{\mathcal{B}_t} \mathbf{v}\, \mathrm{d}m \tag{2.33}$$

such that

$$\mathbf{l}^{\bullet} = \int_{\mathcal{B}_t} \mathbf{a}\, \mathrm{d}m \tag{2.34}$$

and the moment of momentum with respect to some fixed reference point O

$$\mathbf{d}_O = \int_{\mathcal{B}_t} \mathbf{x}_O \times \mathbf{v}\, \mathrm{d}m \tag{2.35}$$

such that

$$\mathbf{d}_O{}^{\bullet} = \int_{\mathcal{B}_t} \mathbf{x}_O \times \mathbf{a}\, \mathrm{d}m. \tag{2.36}$$

On the other side of these Eqs. (2.31) and (2.32), we have the dynamical quantities

$\mathbf{f}$ the resultant force vector

$\mathbf{m}_O$ the resultant torque vector with respect to the reference point O.

In classical continuum mechanics, it is assumed that these two quantities result from distributed forces which act in the interior of the body and on its surface

$$\mathbf{f} = \int_{\mathcal{B}_t} \mathbf{b}\,\mathrm{d}m + \int_{\partial\mathcal{B}_t} \mathbf{t}\,\mathrm{d}A \tag{2.37}$$

$$\mathbf{m}_O = \int_{\mathcal{B}_t} \mathbf{x}_O \times \mathbf{b}\,\mathrm{d}m + \int_{\partial\mathcal{B}_t} \mathbf{x}_O \times \mathbf{t}\,\mathrm{d}A. \tag{2.38}$$

Distributed torques have not been considered in the classical theory so that the torques result exclusively from the distributed forces.

With some additional assumptions, Cauchy could then show by his famous tetrahedron argument that $\mathbf{t}$ is a linear function of the outer normal $\mathbf{n}$ on the surface, which gives rise to introduce the stress tensor $\mathbf{T}$ for

$$\mathbf{t} = \mathbf{T} \cdot \mathbf{n}. \tag{2.39}$$

By some partial integration which can be found in every textbook on continuum mechanics, the balance of linear momentum leads to the field equation

$$\mathrm{div}\mathbf{T} + \rho\,\mathbf{b} = \rho\,\mathbf{a} \tag{2.40}$$

while the balance of moment of momentum requires the symmetry of this stress tensor.

By the introduction of some test function $\delta\mathbf{v}$ called **virtual velocity**, one can further show that the two balance laws hold if and only if

$$\int_{\mathcal{B}_t} \mathbf{b} \cdot \delta\mathbf{v}\,\mathrm{d}m + \int_{\partial\mathcal{B}_t} \mathbf{t} \cdot \delta\mathbf{v}\,\mathrm{d}A = \int_{\mathcal{B}_t} \mathbf{a} \cdot \delta\mathbf{v}\mathrm{d}m + \int_{\mathcal{B}_t} \mathbf{T} \cdot\cdot\, \delta\mathbf{D}\,\mathrm{d}V \tag{2.41}$$

with $\delta\,\mathbf{D} := 1/2\big(\mathrm{grad}\delta\mathbf{v} + \mathrm{grad}^T\delta\mathbf{v}\big)$ holds for all vector fields $\delta\mathbf{v}$. This statement is the ***principle of virtual power***.

Because of this equivalence between the two balances and this principle, one can also start from the *principle of virtual power* and derive the concept of force and torque and their balance laws from it. This may seem to be more elegant since we have to deal only with one scalar equation instead of two vectorial ones. In all cases, one must have a clear knowledge of the forces and torques acting in the interior and on the surface of the body.

Balance Laws for Gradient Materials

For classical or simple materials, it is assumed that the stresses depend on the present and past value of the deformation gradient at this particular point. The intention of gradient models is to extend this dependence to also include higher gradients. In such cases, some fundamental questions arise, such as the following ones.

- Do we also have to invent higher stress tensors?
- If this is so, then what are their symmetry properties?
- And what are their transformation properties under changes of observer?
- Do we also have to invent additional balance laws for them?
- What are the additional boundary conditions for the higher-order kinematical and dynamical quantities?

These are rather fundamental questions, demanding to over-think the very basic assumptions of such a continuum theory.

The first question has been clearly answered by Gurtin (1965) by showing that without additional higher-order stress tensors the second law of thermodynamics would be violated. And it has been shown in the sequel that this shortcoming can be avoided in fact by introducing higher stress tensors of precisely the same order as the higher gradients and with the same symmetry properties. So, there is a certain duality in such theories between the kinematical and the dynamical concepts.

But how to approach the other questions after the form of the balance laws and the boundary conditions in the presence of higher gradients?

Of course, one can try to extend Cauchy's balance laws by higher-order terms. But how do they look like? Without this a priori knowledge of them, we will hardly be able to establish a consistent gradient theory.

The same problem occurs again if we start with the *principle of virtual power*. This has been done by Hellinger (1914) and many others in the sequel like Truesdell and Toupin (1960) and Germain (1972, 1973a, b). For such an approach, one also has to add additional terms on both sides of the equation, the form of which has to be fixed from the outset.

Moreover, this approach has the disadvantage to choose a *virtual*, not a real concept as the starting point. Under philosophical aspects, it may appear questionable to start with an equation which cannot be confirmed by any observation, in principle. More precisely, we can only test this equation for one field of $\delta\mathbf{v}$, namely the current velocity field $\mathbf{v}$. Clearly, a function defined in the infinite-dimensional space of all vector fields is not determined if we know its value for just one element.

For all of these reasons in what follows, we will present a third approach which avoids all such problems (see Bertram and Forest 2007). The steps for this new approach are as follows.

(1) We start with the concept of the (total) power of a body during a certain motion.
(2) We assume that this power is invariant under all changes of observer.
(3) We extend this power in a linear and continuous way to obtain the virtual power functional.

(4) We derive the *principle of virtual power*.
(5) We derive the concepts of forces and torques for this class of materials.
(6) We derive all balance laws and boundary conditions.

This is a long program, which can be reported here only in brief. A detailed derivation can be found in Bertram (2015, 2016, 2017, 2019).

Objectivity of the power functional. The first assumption which we make is that the current power of a body after having passed a particular motion in space can be given as a functional.

Axiom 1 *For each body, there exists a power functional Π with respect to some observer that assigns to any motion of the body $\boldsymbol{\chi}$ the (total)* **power** *$\Pi(\boldsymbol{\chi})$ that the body currently (i.e., at the end of this motion) produces.*
It is zero, whenever the current velocity field is zero for all points of the body.

Here, we wrote for brevity $\boldsymbol{\chi}$ for the motion of the body during a certain interval of time. So, the power may depend on the past or even remote past values of the motion, which is typical for non-elastic effects. But it will in all cases depend on the current kinematical values like the current velocity field at the end of this time interval.

The function $\boldsymbol{\chi}$ for the motion of the body is a purely kinematical concept. The question arises whether some motion is possible for the body in the given surrounding. Traditionally, one would say that this is the case only if all balance laws and boundary conditions are fulfilled. At this stage, however, we do not yet know the balance laws. Therefore, we need another criterion.

We have already seen that the description of the motion depends on the observer, and so does consequently the power functional. The value of this functional, however, shall be invariant under change of observer. This is assumed by

Axiom 2 *A motion of the body is possible if and only if the power is objective under all changes of observer*

$$\Pi^*(\boldsymbol{\chi}^*) = \Pi(\boldsymbol{\chi}) \tag{2.42}$$

at all times during the motion.

We will next consider more precisely the dependence of the power on the current velocity field. For a point force $\mathbf{f}$, the power will typically be of the form

$$\Pi(\boldsymbol{\chi}) = \mathbf{f} \cdot \mathbf{v} \tag{2.43}$$

so one is tempted to say that the power is linear in the velocity. However, one cannot check a function for linearity if its value is only known for one argument, namely $\mathbf{v}$. What we can do is to extend such a function to other than the current velocity by writing $\mathbf{f} \cdot \delta\mathbf{v}$ which would correspond to the previously introduced virtual velocity. We will now extend this procedure to our three-dimensional body.

For that purpose, we introduce the *space of all differentiable vector fields* on the body by $\delta\mathcal{V}$. One prominent member of this space would always be the current velocity field $\mathbf{v}$.

This space is a vector space for which the linear operations are introduced pointwise, as usual. The topological structure, however, is non-trivial, since it is a function space (with infinite dimension). We will assume further on that all virtual velocity fields are K-times piecewise differentiable in space for some $K \geq 0$, which shall be specified later and specifies the order of the theory. A topological structure on $\delta\mathcal{V}$ is introduced by the Sobolev $(2,K)$-norm

$$|\delta\mathbf{v}|^K := \sqrt{\int_{B_t} \left[|\delta\mathbf{v}(\mathbf{x})|^2 + |\text{grad}\,\delta\mathbf{v}(\mathbf{x})|^2 + \ldots + \left|\text{grad}^K\delta\mathbf{v}(\mathbf{x})\right|^2\right] dV} \tag{2.44}$$

for all $\delta\mathbf{v} \in \delta\mathcal{V}$. Here, grad^K means the K-fold Eulerian gradient of the field $\delta\mathbf{v}$

$$\text{grad}^K\delta\mathbf{v} := \delta\mathbf{v} \otimes \nabla^K = \delta\mathbf{v} \otimes \nabla \otimes \ldots \otimes \nabla (K - \text{times}). \tag{2.45}$$

This norm makes a topological vector space out of $\delta\mathcal{V}$, which is contained in the usual Sobolev[1] space $\mathcal{W}^{2,K}$.

We are now able to introduce the *virtual power* as a continuous extension of the *power* being linear in the virtual velocities.

Definition. For a given motion $\boldsymbol{\chi}$ of a body with respect to some observer, the **virtual power functional** $\delta\Pi(\boldsymbol{\chi}, \delta\mathbf{v})$ depends on the motion and on the virtual velocity field such that for each motion

$$\delta\Pi(\boldsymbol{\chi}, \bullet) : \delta\mathcal{V} \to \mathcal{R}$$

(P1) is continuous and linear.

(P2) extends the power functional, *i.e.*,

$$\delta\Pi(\boldsymbol{\chi}, \mathbf{v}) = \Pi(\boldsymbol{\chi}) \tag{2.46}$$

holds for the current velocity field $\mathbf{v}$.

(P3) transforms like the power functional, *i.e.*, for all observers we have

$$\delta\Pi^*(\boldsymbol{\chi}^*, \delta\mathbf{v}^*) - \delta\Pi(\boldsymbol{\chi}, \delta\mathbf{v}) = \Pi^*(\boldsymbol{\chi}^*) - \Pi(\boldsymbol{\chi}) \tag{2.47}$$

if $\boldsymbol{\chi}$ is transformed after (2.20) and $\delta\mathbf{v}$ are transformed analogously as

[1]See Adams (1975).

$$\delta\mathbf{v}^* = \mathbf{Q}\cdot\delta\mathbf{v} + \mathbf{Q}^\bullet\cdot\mathbf{Q}^T\cdot\left(\mathbf{x}^* - \mathbf{c}\right) + \mathbf{c}^\bullet. \tag{2.48}$$

Two questions immediately arise. First, does such an extension always exist, and second, if so then is it unique?

The existence cannot be proven generally. We can only state that for all known examples it does exist. If this is so, one can easily see that it cannot be unique. Instead, infinitely many such extensions exist. We have to keep this in mind when we draw conclusions from the virtual power functional.

If we substitute (2.48) into (2.47) and use the linearity of the virtual power functional, we obtain

$$\begin{aligned}&\delta\Pi^*\left(\chi^*, \mathbf{Q}\cdot\delta\mathbf{v}\right) - \delta\Pi(\chi, \delta\mathbf{v})\\ &= \Pi^*\left(\chi^*\right) - \Pi(\chi) - \delta\Pi^*\left(\chi^*, \mathbf{Q}^\bullet\cdot\mathbf{Q}^T\cdot\left(\mathbf{x}^* - \mathbf{c}\right)\right) - \delta\Pi^*\left(\chi^*, \mathbf{c}^\bullet\right)\end{aligned} \tag{2.49}$$

for all $\delta\mathbf{v} \in \delta\mathcal{V}$. Regarding the dependencies upon $\delta\mathbf{v}$, the right-hand side of this equation is constant. The only linear function that equals a constant is the zero function. Thus,

$$\delta\Pi^*\left(\chi^*, \mathbf{Q}\cdot\delta\mathbf{v}\right) = \delta\Pi(\chi, \delta\mathbf{v}) \quad \text{for all} \quad \delta\mathbf{v} \in \delta\mathcal{V}. \tag{2.50}$$

This equality will be later used to demonstrate the objectivity of the stress tensors.

Next, we consider the virtual power functional for two rather special virtual velocity fields occurring on the left-hand side of (2.49), namely those which result from a constant field $\delta\mathbf{v}_O \in \delta\mathcal{V}$ and a rotational field $\delta\boldsymbol{\omega}\times\mathbf{x} \in \delta\mathcal{V}$ with two constant vectors $\delta\mathbf{v}_O$ and $\delta\boldsymbol{\omega}$.

Because of the linearity of the virtual power, there exist two dual vectors $\underline{\mathbf{f}}$ and $\underline{\mathbf{m}}_O$ such that

$$\delta\Pi(\chi, \delta\mathbf{v}_O + \delta\boldsymbol{\omega}\times\mathbf{x}) = \underline{\mathbf{f}}\cdot\delta\mathbf{v}_O + \underline{\mathbf{m}}_O\cdot\delta\boldsymbol{\omega} \tag{2.51}$$

holds for all constant vectors $\delta\mathbf{v}_O$ and $\delta\boldsymbol{\omega}$. We will call

$\underline{\mathbf{f}}$ the generalized force, and
$\underline{\mathbf{m}}_O$ the generalized torque with respect to O.

Since the power depends in general on the motion of the body, this holds also for the generalized forces and torques.

These two dynamical quantities still contain inertial contributions. We define therefore the **force**

$$\mathbf{f} := \underline{\mathbf{f}} + \mathbf{l}^\bullet \tag{2.52}$$

and the **torque** with respect to O

$$\mathbf{m}_O := \underline{\mathbf{m}}_O + \mathbf{d}_O^\bullet \tag{2.53}$$

with the linear momentum $\mathbf{l}$ from (2.33) and the angular momentum $\mathbf{d}_O$ from (2.35). Both of these quantities are global ones. They are defined for the body at a certain instant of time.

With these definitions, it is not difficult to show that the power functional is objective if and only if the following equivalent conditions hold.

- The virtual power is zero for vector fields of the form $\delta\mathbf{v}_O + \delta\boldsymbol{\omega} \times \mathbf{x}$ with two arbitrary vectors $\delta\mathbf{v}_O$ and $\delta\boldsymbol{\omega}$ (*principle of virtual power*)

$$\delta\Pi(\chi, \delta\mathbf{v}_O + \delta\boldsymbol{\omega} \times \mathbf{x}) = 0 \tag{2.54}$$

- The generalized forces and torques are in equilibrium (*extended principle of d'Alembert*),

$$\underline{\mathbf{f}} = \mathbf{o} \quad \text{equilibrium condition of generalized forces} \tag{2.55}$$

$$\underline{\mathbf{m}}_O = \mathbf{o} \quad \text{equilibrium condition of generalized torques} \tag{2.56}$$

with respect to some/any reference point O

- The two balance laws hold,

$$\mathbf{l}^\bullet = \mathbf{f} \quad \text{balance of linear momentum} \tag{2.57}$$

$$\mathbf{d}_O{}^\bullet = \mathbf{m}_O \quad \text{balance of moment of momentum} \tag{2.58}$$

with respect to some/any reference point O

Field formulation. In order to obtain a field formulation of the dynamics of gradient materials, we apply the Riesz representation theorem of linear continuous functionals on topological vector spaces.

Theorem *For each observer, there exist K + 1 time-dependent tensor fields* $\overset{\langle i \rangle}{\boldsymbol{T}}$ *of order i = 1, ..., K + 1 such that*

$$\delta\Pi(\chi, \delta\mathbf{v}) = \int_{\mathcal{B}_t} \Big(\overset{\langle 1 \rangle}{\boldsymbol{T}} \cdot \delta\mathbf{v} + \overset{\langle 2 \rangle}{\boldsymbol{T}} \cdot\cdot \operatorname{grad} \delta\mathbf{v} + \ldots$$

$$+ \overset{\langle K+1\rangle}{\boldsymbol{T}} \cdot \ldots \cdot \operatorname{grad}^K \delta\mathbf{v} \Big) dV \quad \forall \delta\mathbf{v} \in \delta\mathcal{V}. \tag{2.59}$$

The tensors $\overset{\langle 3\rangle}{\boldsymbol{T}} \ldots \overset{\langle K+1\rangle}{\boldsymbol{T}}$ are called **hyperstresses**. In general, they are again functionals of the motion of the body.

Since the higher velocity gradients show the right subsymmetries, we can also assume the same subsymmetries for the hyperstresses, like

$$\begin{aligned} \overset{\langle 3\rangle}{T}_{ijk} &= \overset{\langle 3\rangle}{T}_{ikj} \\ \overset{\langle 4\rangle}{T}_{ijkl} &= \overset{\langle 4\rangle}{T}_{ikjl} = \overset{\langle 4\rangle}{T}_{ijlk} \end{aligned} \tag{2.60}$$

etc.

By (2.47), we obtain the same representation for the power

$$\Pi(\chi) = \int_{\mathcal{B}_t} \left(\overset{\langle 1\rangle}{\boldsymbol{T}} \cdot \mathbf{v} + \overset{\langle 2\rangle}{\boldsymbol{T}} \cdot\cdot \operatorname{grad} \mathbf{v} + \ldots + \overset{\langle K+1\rangle}{\boldsymbol{T}} \cdot \ldots \cdot \operatorname{grad}^K \mathbf{v} \right) dV. \tag{2.61}$$

By substituting (2.59) into (2.50), we see that these dynamical fields are objective under changes of observers

$$\overset{\langle i\rangle}{\boldsymbol{T}} = \mathbf{Q} * \overset{\langle i\rangle}{\boldsymbol{T}} \quad \text{for } i = 1, \ldots, K+1. \tag{2.62}$$

In the rest of this chapter, we particularize our focus on the following

Second-order gradient materials. The second term of (2.59) is already familiar to us from the classical theory of simple materials. It can be transformed by partial integration into

$$\begin{aligned} &\int_{\mathcal{B}_t} \overset{\langle 2\rangle}{\boldsymbol{T}} \cdot\cdot \operatorname{grad} \delta\mathbf{v} \, dV \\ &= \int_{\partial\mathcal{B}_t} (\overset{\langle 2\rangle}{\boldsymbol{T}} \cdot \mathbf{n}) \cdot \delta\mathbf{v} \, dA - \int_{\mathcal{B}_t} \operatorname{div} \overset{\langle 2\rangle}{\boldsymbol{T}} \cdot \delta\mathbf{v} \, dV \end{aligned} \tag{2.63}$$

with the outer normal $\mathbf{n}$ on the surface of the body.

The third term is not present in classical theories. It can be twice integrated by parts. Firstly,

$$\int_{\mathcal{B}_t} \overset{\langle 3\rangle}{\boldsymbol{T}} \therefore \operatorname{grad}^2 \delta\mathbf{v} \, dV$$

$$= \int_{\partial \mathcal{B}_t} (\overset{\langle 3 \rangle}{\boldsymbol{T}} \cdot \mathbf{n}) \cdot \cdot \operatorname{grad} \delta \mathbf{v} \, dA - \int_{\mathcal{B}_t} \operatorname{div} \overset{\langle 3 \rangle}{\boldsymbol{T}} \cdot \cdot \operatorname{grad} \delta \mathbf{v} \, dV. \tag{2.64}$$

If $\mathcal{B}_t$ denotes the region that the body currently occupies in space, and $\partial \mathcal{B}_t$ its boundary, this may be subdivided into a finite set of surface segments, bounded by edges, which are denoted by $\mathcal{L}_t$. In order to avoid sums and indices, the expression $\int_{\mathcal{L}_t}$ stands for the line integrals over *all* edges of *all* surface segments. So, every edge line enters twice since it belongs to two adjacent surface regions. We introduce a Frenet tripod on the borderline with respect to one of the adjacent surface elements.

$\mathbf{t}_{\mathcal{L}}$ Frenet's tangent to the edge line
$\mathbf{n}$ outer normal to the surface segments
$\mathbf{m} := \mathbf{t}_{\mathcal{L}} \times \mathbf{n}$

so that $\{ \mathbf{t}_{\mathcal{L}}, \mathbf{n}, \mathbf{m}\}$ forms a positive-oriented orthonormal vector base.

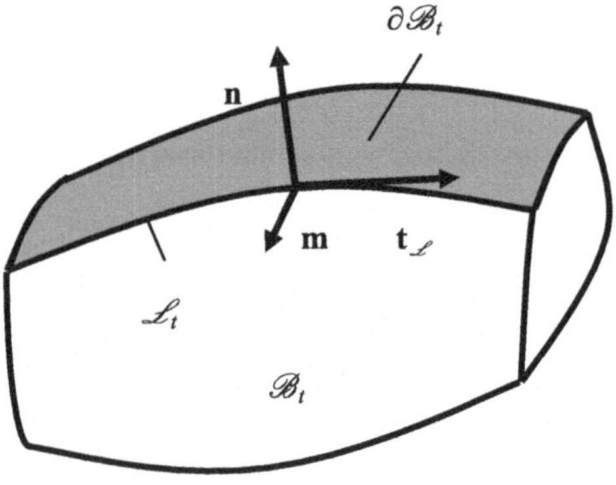

We decompose the gradient of a differentiable function φ at surface points into its normal and its tangential part

$$\operatorname{grad}\varphi = \operatorname{grad}_n \varphi + \operatorname{grad}_t \varphi \tag{2.65}$$

which corresponds to the natural split of the spatial nabla operator $\boldsymbol{\nabla} = \boldsymbol{\nabla}_n + \boldsymbol{\nabla}_t$ with its normal part

$$\nabla_n := \mathbf{n} \otimes \mathbf{n} \cdot \nabla = \nabla \cdot \mathbf{n} \otimes \mathbf{n} = \frac{\partial}{\partial x_n} \mathbf{n} \tag{2.66}$$

with the outer normal $\mathbf{n}$ to the surface and the normal coordinate x_n, and the tangential part

$$\nabla_t := \nabla \cdot (\mathbf{I} - \mathbf{n} \otimes \mathbf{n}). \tag{2.67}$$

The trace of these gradients is the divergence which is also decomposed

$$\operatorname{div}\varphi = \operatorname{div}_n\varphi + \operatorname{div}_t\varphi \tag{2.68}$$

If we decompose the gradient within the following expression with a vector field $\mathbf{v}$ and a second-order tensor field $\overset{\langle 2\rangle}{\boldsymbol{T}}$ and use the product rule, we obtain

$$\begin{aligned}\overset{\langle 2\rangle}{\boldsymbol{T}} \cdot\cdot \operatorname{grad}\mathbf{v} &= \overset{\langle 2\rangle}{\boldsymbol{T}} \cdot\cdot \operatorname{grad}_n\mathbf{v} + \overset{\langle 2\rangle}{\boldsymbol{T}} \cdot\cdot \operatorname{grad}_t\mathbf{v}\\ &= \overset{\langle 2\rangle}{\boldsymbol{T}} \cdot\cdot \operatorname{grad}_n\mathbf{v} + \operatorname{div}_t(\mathbf{v}\cdot\overset{\langle 2\rangle}{\boldsymbol{T}}) - \left(\operatorname{div}_t\overset{\langle 2\rangle}{\boldsymbol{T}}\right)\cdot\mathbf{v}.\end{aligned} \tag{2.69}$$

The surface divergence theorem holds in the form

$$\int\limits_{\partial\mathcal{B}_t} \operatorname{div}_t\mathbf{v}\,\mathrm{d}A = \int\limits_{\partial\mathcal{B}_t} (\operatorname{div}_t\mathbf{n})\mathbf{v}\cdot\mathbf{n}\,\mathrm{d}A + \int\limits_{\mathcal{L}_t} \mathbf{v}\cdot\mathbf{m}\,\mathrm{d}L. \tag{2.70}$$

With this, we can further integrate by parts

$$\begin{aligned}&\int\limits_{\mathcal{B}_t} \overset{\langle 3\rangle}{\boldsymbol{T}} \therefore \operatorname{grad}^2\delta\mathbf{v}\,\mathrm{d}V\\ &= \int\limits_{\partial\mathcal{B}_t}\left[(\overset{\langle 3\rangle}{\boldsymbol{T}}\cdot\mathbf{n})\cdot\cdot\operatorname{grad}_n\delta\mathbf{v} - \left(\operatorname{div}\overset{\langle 3\rangle}{\boldsymbol{T}}\cdot\mathbf{n}\right)\cdot\delta\mathbf{v}\right]\mathrm{d}A + \int\limits_{\mathcal{B}_t}\operatorname{div}^2\overset{\langle 3\rangle}{\boldsymbol{T}}\cdot\delta\mathbf{v}\,\mathrm{d}V\\ &+ \int\limits_{\partial\mathcal{B}_t}\left[\overset{\langle 3\rangle}{\boldsymbol{T}}\cdot\cdot\left(\operatorname{div}_t\mathbf{n}\,\mathbf{n}\otimes\mathbf{n} - \operatorname{grad}_t\mathbf{n}\right) - \operatorname{div}_t\overset{\langle 3\rangle}{\boldsymbol{T}}\cdot\mathbf{n}\right]\cdot\delta\mathbf{v}\,\mathrm{d}A\\ &+ \int\limits_{\mathcal{L}_t}(\overset{\langle 3\rangle}{\boldsymbol{T}}\cdot\cdot\mathbf{m}\otimes\mathbf{n})\cdot\delta\mathbf{v}\,\mathrm{d}L.\end{aligned} \tag{2.71}$$

If we substitute these results into the virtual power, then we obtain for a second-order gradient material

$$\begin{aligned}&\int\limits_{\mathcal{B}_t}(\mathbf{b}-\mathbf{a})\cdot\delta\mathbf{v}\,\mathrm{d}m + \int\limits_{\mathcal{B}_t}\left[\mathbf{t}_{\mathcal{A}}\cdot\delta\mathbf{v} + (\overset{\langle 3\rangle}{\boldsymbol{T}}\cdot\mathbf{n})\cdot\cdot\operatorname{grad}_n\delta\mathbf{v}\right]\mathrm{d}A\\ &+ \int\limits_{\mathcal{L}_t}\mathbf{t}_{\mathcal{L}}\cdot\delta\mathbf{v}\,\mathrm{d}L = \int\limits_{\partial\mathcal{B}_t}(\overset{\langle 2\rangle}{\boldsymbol{T}}\cdot\cdot\,\delta\mathbf{D} + \overset{\langle 3\rangle}{\boldsymbol{T}}\cdot\cdot\cdot\operatorname{grad}\operatorname{grad}\delta\mathbf{v})\,\mathrm{d}V\end{aligned} \tag{2.72}$$

with the following abbreviations

- the virtual rate of stretching tensor

$$\delta\mathbf{D} := sym\operatorname{grad}\delta\mathbf{v} \tag{2.73}$$

- the body force per unit volume

$$\rho\,\mathbf{b} := \mathbf{t} - \operatorname{div}\overset{\langle 2\rangle}{\boldsymbol{T}} + \operatorname{div}^2\overset{\langle 3\rangle}{\boldsymbol{T}} + \rho\,\mathbf{a} \tag{2.74}$$

- the surface tension vector

$$\begin{aligned}\mathbf{t}_{\mathcal{A}} := {} & (\overset{\langle 2\rangle}{\boldsymbol{T}} - \operatorname{div}_n\overset{\langle 3\rangle}{\boldsymbol{T}} - 2\operatorname{div}_t\overset{\langle 3\rangle}{\boldsymbol{T}})\cdot\mathbf{n} \\ & + \overset{\langle 3\rangle}{\boldsymbol{T}}\cdot\cdot(\operatorname{div}_t\mathbf{n}\,\mathbf{n}\otimes\mathbf{n} - \operatorname{grad}_t\mathbf{n})\end{aligned} \tag{2.75}$$

- the specific line force on the edge

$$\mathbf{t}_{\mathcal{L}} := \overset{\langle 3\rangle}{\boldsymbol{T}}\cdot\cdot\,\mathbf{m}\otimes\mathbf{n} \tag{2.76}$$

so that the **balance of linear momentum** is

$$\int_{\mathcal{B}_t}\mathbf{b}\,\mathrm{d}m + \int_{\partial\mathcal{B}_t}\mathbf{t}_{\mathcal{A}}\;\mathrm{d}A + \int_{\mathcal{L}_t}\mathbf{t}_{\mathcal{L}}\;\mathrm{d}L = \int_{\mathcal{B}_t}\mathbf{a}\,\mathrm{d}m \tag{2.77}$$

and the **balance of moment of momentum**

$$\begin{aligned}&\int_{\mathcal{B}_t}\mathbf{x}\times\mathbf{b}\,\mathrm{d}m + \int_{\partial\mathcal{B}_t}\left[\mathbf{x}\times\mathbf{t}_{\mathcal{A}} + 2\,axi_n(\overset{\langle 3\rangle}{\boldsymbol{T}}\cdot\mathbf{n})\right]\mathrm{d}A + \int_{\mathcal{L}_t}\mathbf{x}\times\mathbf{t}_{\mathcal{L}}\;\mathrm{d}L \\ &= \int_{\mathcal{B}_t}\mathbf{x}\times\mathbf{a}\,\mathrm{d}m\end{aligned} \tag{2.78}$$

with the normal axial vector $axi_n(\overset{\langle 3\rangle}{\boldsymbol{T}}\cdot\mathbf{n})$ of the second-order tensor $\overset{\langle 3\rangle}{\boldsymbol{T}}\cdot\mathbf{n}$ defined by

$$axi_n(\overset{\langle 3\rangle}{\boldsymbol{T}}\cdot\mathbf{n})\cdot\boldsymbol{\omega} := 1/2(\overset{\langle 3\rangle}{\boldsymbol{T}}\cdot\mathbf{n})\cdot\operatorname{grad}_n(\boldsymbol{\omega}\times\mathbf{x}) \tag{2.79}$$

for any vector $\boldsymbol{\omega}$.

If one prefers field equations instead of these global balances, one obtains the *extended Cauchy's laws*

$$\operatorname{div}\left(\overset{\langle 2\rangle}{\boldsymbol{T}} - \operatorname{div}\overset{\langle 3\rangle}{\boldsymbol{T}}\right) + \rho\,\mathbf{b} = \rho\,\mathbf{a} \tag{2.80}$$

$$\overset{\langle 2\rangle}{\boldsymbol{T}} = \overset{\langle 2\rangle}{\boldsymbol{T}}{}^{T} \tag{2.81}$$

to hold everywhere in the body.

From the above balances, we can also see the boundary conditions appropriate for a second gradient theory.

The dynamic or **Neumann boundary conditions** for the body are

- the vector field of the tractions $\mathbf{t}_{\mathcal{A}}$ on $\partial\mathcal{B}_t$ working on $\delta\mathbf{v}$
- the line forces $\mathbf{t}_{\mathcal{L}}$ on edges $\mathcal{L}_t$ working on $\delta\mathbf{v}$
- and the tensor field of the double tractions $\overset{\langle 3\rangle}{\boldsymbol{T}} \cdot \mathbf{n}$ in normal direction on $\partial\mathcal{B}_t$ working on $\text{grad}_n\, \delta\mathbf{v}$.

The kinematical or **Dirichlet boundary conditions** are then the prescription of

- the displacement field $\mathbf{u}$ on the surface of the body
- its normal gradient $\text{grad}_n\, \mathbf{u}$ on the surface of the body
- and the displacements of the edges (in a compatible way).

The analogous analysis for third-order gradients has been done by Bertram (2015, 2017, 2019).

We can observe that the step from a simple material to a second gradient material is neither simple nor straightforward. And one can hardly extrapolate them to higher gradients without making lengthy calculations.

Only Cauchy's laws can be easily extended to an Nth-order gradient material without problems; namely, the balance of linear momentum becomes

$$\text{div}\,\overset{\langle 2\rangle}{\boldsymbol{T}} - \text{div}^2\,\overset{\langle 3\rangle}{\boldsymbol{T}} + \text{div}^3\,\overset{\langle 4\rangle}{\boldsymbol{T}} - \ldots + (-1)^{N+1}\text{div}^N\,\overset{\langle N+1\rangle}{\boldsymbol{T}} + \rho\,\mathbf{b} = \rho\,\mathbf{a} \tag{2.82}$$

while the balance of angular momentum remains the same as in (2.81).

References

Adams, R. A. (1975). *Sobolev spaces*. Academic Press.

Bertram, A., & Forest, S. (2007). Mechanics based on an objective power functional. *Techn. Mechanik, 27*(1), 1–17.

Bertram, A. (2015, 2016, 2017, 2019) *Compendium on Gradient Materials* https://www.lkm.tu-berlin.de/menue/mitarbeiterinnen_sprechstunden/prof_dr_ing_a_bertram/.

Germain, P. (1972). Sur l'application de la méthode des puissances virtuelles en mécanique des milieux continus. *C.R. Acad. Sci. Paris Sér. A, 274*(2), 1051–1055.

Germain, P. (1973a). La méthode des puissances virtuelles en mécanique des milieux continus, première partie: théorie du second gradient. *Journal of Mécanique, 12,* 235–274.

Germain, P. (1973b). The method of virtual power in continuum mechanics. Part 2: Microstructure. *Journal of Applied Mathematics, 25,* 556–575.

Gurtin, M. E. (1965). Thermodynamics and the possibility of spacial interaction in elastic materials. *Archive for Rational Mechanics and Analysis, 19*(5), 339–352.

Hellinger, E. (1914). Die allgemeinen Ansätze der Mechanik der Kontinua. In: *Encyklopädie der mathematischen Wissenschaften, IV*(30), 601–694.

Truesdell, C., & Toupin, R. A. (1960). *The classical field theories.* In: S. Flügge (Ed.), *Handbuch der Physik* (Vol. III/1). Springer, Berlin.

Chapter 3
Strain Gradient Elasticity From Capillarity to the Mechanics of Nano-objects

Samuel Forest

Abstract This chapter provides examples of the application of strain gradient elasticity theory. It starts with the consideration of capillary effects in elastic fluids and its representation by Mindlin's second strain gradient theory. The same model is used to simulate size effects in the mechanical response of nano-wires and nano-films represented by linear isotropic second strain gradient media. The relevance of the strain gradient elasticity approach is then discussed for the modeling of the dispersion of elastic waves. The chapter ends with a comparison of the strain gradient model with the recently developed stress gradient continuum. Fundamental differences are put forward, and an example is provided showing that stress gradient elasticity predicts the existence of a boundary layer at the free surfaces of a specimen under tensile loading.

Keywords Apparent shear modulus · Capillarity · Gradient fluid · Nano-objects · Strain gradient elasticity · Stress gradient elasticity

Capillarity in Elastic Strain Gradient Fluids

Due to their different local environments, atoms at a free surface and atoms in the bulk of a material have different associated energy and lattice spacing. This excess of energy associated with surface atoms is called surface free energy and gives rise to the surface tension. The surface region is only a few atomic layers thin that is why the surface tension can be neglected when the characteristic length of the microstructure of the considered material is in the micrometer range or larger. But in the case of nano-sized materials, the ratio between the surface and the volume is much higher and the surface region behavior cannot be neglected anymore. There are several ways to introduce the mechanical properties of the surface. Two homogeneous phases

S. Forest (✉)
MINES ParisTech, PSL University, Centre des matériaux (CMAT), CNRS UMR 7633, 87 91003 Evry, France
e-mail: samuel.forest@mines-paristech.fr

A. Bertram and S. Forest (eds.), *Mechanics of Strain Gradient Materials*, CISM International Centre for Mechanical Sciences 600,
https://doi.org/10.1007/978-3-030-43830-2_3

can be considered as separated by a single dividing surface. The thermodynamic properties of the interface are defined as the excess over the values obtained for both bulk phases separated by a zero-thickness surface (Müller and Saúl 2004; Dingreville et al. 2005). The continuum mechanical theory of surface/interface behavior has been settled by Gurtin and Murdoch (1978) and examined recently by Javili et al. (2018). It introduces a volume stress tensor in the bulk of the material and a surface stress tensor in the surface or interface modeled as a membrane. Both stress tensors fulfill the balance of momentum equations. A specific elastic behavior is attributed to the membrane, and kinematic constraints ensure that the bulk part and the surface remain coherent. The most common manifestation of surface behavior is capillarity effects in elastic fluids. It is described by the Young–Laplace equation which states that the internal pressure, p, in a spherical droplet is proportional to the surface tension, γ, multiplied by the surface curvature, $1/r$:

$$p = \frac{2\gamma}{r}. \tag{3.1}$$

When the size of the considered object is small enough, there is no clear way to define a sharp interface or surface. Instead, a continuum model can be used to describe a transition domain between two bulk regions, or between the bulk and the outer free surface. Such continuum theories for diffuse surface or interfaces have been developed for a refined description of capillarity in elastic fluids and solids. They are based on higher-order gradient theories like the Korteweg equation which involves the gradient of density; see Truesdell and Noll (1965), Sect. 124. A more general strain gradient theory has been proposed by Casal (1963, 1972) and Casal and Gouin (1985). In his celebrated paper, Mindlin (1965) claims that the second gradient of strain or, equivalently, the third gradient of displacement theory is in fact needed to describe, in a continuous manner, capillarity and cohesion effects in isotropic linear elastic solids and fluids. Based on a simple one-dimensional atomic chain model, he identifies the higher-order elasticity modulus that is responsible for the variation of lattice spacing from the free surface into the bulk in a semi-infinite crystal.

In this section, Korteweg's equation is compared with the first and second gradient of strain theories in order to highlight the main differences. For the sake of conciseness, the small strain framework is adopted. Volume forces are not considered throughout the work. The analysis is limited to the static case. Mindlin's notation is adopted and combined with those from Chap. 1 of this book.

Korteweg's Equation and First Strain Gradient Model

The Van der Waals and Korteweg equations are the first attempts to introduce capillary effects in a continuum mechanical theory. For an elastic medium, they include not only the effect of mass density, ρ, on stress but also of its gradient $\nabla\rho$, in the form

$$\boldsymbol{\tau} = -p(\rho)\mathbf{1} - \alpha(\nabla\rho)^2\mathbf{1} - \beta\nabla\rho\otimes\nabla\rho + \gamma(\nabla^2\rho)\mathbf{1} + \delta\nabla\otimes\nabla\rho, \tag{3.2}$$

where $\mathbf{1}$ is the second-order identity tensor, $\boldsymbol{\tau}$ is a second rank stress tensor and α, β, γ and δ are material parameters, namely higher-order elasticity moduli, ∇^2 is the Laplace operator and

$$\nabla\otimes\nabla\rho = \rho_{,ij}\,\boldsymbol{e}_i\otimes\boldsymbol{e}_j, \tag{3.3}$$

in a Cartesian orthonormal basis. The divergence of $\boldsymbol{\tau}$ is assumed to vanish, in the absence of volume forces. In their account of Korteweg's constitutive theory, Truesdell and Noll (1965) show how it can be used to represent a spherical non-uniform field of mass density, $\rho(r)$, thus allowing for the presence of an interface between liquid and vapor in a water droplet. The balance equation,

$$T'_{rr} + \frac{2}{r}(T_{rr} - T_{\theta\theta}) = 0, \tag{3.4}$$

is combined with the constitutive relation,

$$T_{rr} - T_{\theta\theta} = -\beta\rho'^2 - \delta\frac{\rho'}{r} + \delta\rho'', \tag{3.5}$$

where the prime denotes derivation with respect to r. Integrating equation (3.4) on the interface zone $[r_1, r_2]$ yields the relation

$$T_{rr}(r_2) - T_{rr}(r_1) = 2\delta\left(\frac{\rho'(r_2)}{r_2} - \frac{\rho'(r_1)}{r_1}\right) - 2\int_{r_1}^{r_2}\frac{\beta\rho'^2 - \delta\rho''}{r}dr. \tag{3.6}$$

In order to obtain results appropriate to a thin shell of transition, we calculate the limit of the latter relation as r_1, r_2 tend to r_0. The first term on the right-hand side vanishes, whereas under suitable regularity assumptions, the second term is proportional to the mean curvature, $1/r_0$. Accordingly, Eq. (3.6) can be interpreted as the diffuse counterpart of Laplace's sharp interface equation.

The consistency of such higher gradient constitutive equations, formulated within the framework of classical continuum mechanics, with respect to continuum thermodynamics has been questioned by Gurtin (1965) (see also Rocard 1967). It is argued that such higher-order constitutive statements can be acceptable only if higher-order stress tensors are introduced in addition to the usual Cauchy simple force stress tensor. To see that, let us now rephrase Korteweg's equation within the linear elasticity framework. For that purpose, the dilatation[1] Δ is defined as the trace of the small strain tensor,

$$\Delta = \operatorname{trace}\boldsymbol{\varepsilon} = 1 - \frac{\rho_0}{\rho}, \quad \nabla\Delta = \rho_0\frac{\nabla\rho}{\rho^2} \simeq \frac{\nabla\rho}{\rho_0}, \tag{3.7}$$

[1] Mindlin's notation Δ is used but it should not be mixed with the Laplace operator denoted by ∇^2 in this chapter.

with respect to a reference mass density ρ_0.

The usual strain tensor is $\varepsilon = \frac{1}{2}(\boldsymbol{u} \otimes \nabla + \nabla \otimes \boldsymbol{u})$.

The Korteweg equation (3.2) can therefore be written as

$$\boldsymbol{\tau} = -p(\Delta)\mathbf{1} - \alpha(\nabla\Delta)^2\mathbf{1} - \beta\nabla\Delta \otimes \nabla\Delta + \gamma(\nabla^2\Delta)\mathbf{1} + \delta\nabla \otimes \nabla\Delta, \tag{3.8}$$

$$\tau_{ij} = -p(\Delta)\delta_{ij} - \alpha\Delta_{,k}\Delta_{,k}\delta_{ij} - \beta\Delta_{,i}\Delta_{,j} + \gamma\Delta_{,kk}\delta_{ij} + \delta\Delta_{,ij}. \tag{3.9}$$

with the same elastic moduli as in Eq. (3.2). The linearized Korteweg equation follows as

$$\boldsymbol{\tau} = -p(\Delta)\mathbf{1} + \gamma(\nabla^2\Delta)\mathbf{1} + \delta\nabla \otimes \nabla\Delta. \tag{3.10}$$

It involves higher-order gradients of the dilatation. This suggests that it could be derived from a strain gradient theory as proposed by Toupin (1962), Casal (1963) and Mindlin and Eshel (1968). As shown by Toupin, the second gradient of displacement theory is equivalent to the first strain gradient theory. Such theories introduce higher-order stresses, as advocated by Gurtin (1965). The strain gradient theory can be limited to the gradient of density effects and therefore compared to Korteweg's equation, as done by Casal and Gouin (1985). The relation between the stress tensor $\boldsymbol{\tau}$ in Eq. (3.2) and the higher-order stress measures of strain gradient theories is discussed in the next section. More recent works have developed the concept of a fluid with internal wettability based on modified Korteweg equations within the strain gradient framework, also called Cahn–Hilliard fluid in Seppecher (2000, 1996).

This idea of considering higher gradients of density to describe the material behavior close to free surfaces or interfaces has been implemented for phase transformations like in water droplets in vapor (Dreyer and Duderstadt 2006). This represents a diffuse interface model for liquid–gas interfaces.

Application of Second Strain Gradient Theory

Mindlin (1965) claims that a second strain gradient theory is necessary to account for capillarity and cohesion effects in isotropic elastic media, instead of the first strain gradient theory reported in the previous section. The second strain gradient theory is presented and contains the first strain gradient model as a special case. It is based on the assumption that the stress state at a material point depends on the values of the strain, the first and the second strain gradients at that point. According to the results of chapter "The Experimental Evidence for Higher Gradient Theories," the work of internal forces of a second strain gradient media is

$$w^{(i)} = \overset{2}{\boldsymbol{T}} : \varepsilon + \overset{3}{\boldsymbol{T}} \,\vdots\, \overset{3}{\varepsilon} + \overset{4}{\boldsymbol{T}} :: \overset{4}{\varepsilon} \tag{3.11}$$

where $\overset{2}{\boldsymbol{T}}$ is the simple force stress tensor of second rank and $\overset{3}{\boldsymbol{T}}, \overset{4}{\boldsymbol{T}}$ are the third and fourth rank hyperstress tensors conjugate to the strain tensors

$$\varepsilon_{ij} = u_{(i,j)}, \quad \varepsilon_{ijk} = \varepsilon_{ij,k}, \quad \varepsilon_{ijkl} = \varepsilon_{ij,kl} \tag{3.12}$$

As established in chapter "The Experimental Evidence for Higher Gradient Theories," the generalized stress tensors must fulfill the following balance equation

$$\boldsymbol{\tau} \cdot \nabla = 0, \quad \text{with} \quad \boldsymbol{\tau} = \overset{2}{\boldsymbol{T}} - (\overset{3}{\boldsymbol{T}} - \overset{4}{\boldsymbol{T}} \cdot \nabla) \cdot \nabla \tag{3.13}$$

The stress tensor $\boldsymbol{\tau}$ is an effective stress tensor whose divergence vanishes. It is expressed in terms of the stress tensors of all orders. This partial differential equation is accompanied by three sets of complex boundary conditions that involve the surface curvature and the normal and tangent derivatives of stress quantities. For the sake of conciseness, they are not recalled here and the author is referred to Eqs. (18a, b, c) in Mindlin (1965). Corresponding expressions of the boundary conditions for the first gradient of strain theory can be found in Mindlin and Eshel (1968), Germain (1973a). The free energy density function, $\Psi(\varepsilon, \overset{3}{\varepsilon}, \overset{4}{\varepsilon})$, is then a potential from which stresses are derived:

$$\overset{2}{\boldsymbol{T}} = \rho \frac{\partial \Psi}{\partial \varepsilon}, \quad \overset{3}{\boldsymbol{T}} = \rho \frac{\partial \Psi}{\partial \overset{3}{\varepsilon}}, \quad \overset{4}{\boldsymbol{T}} = \rho \frac{\partial \Psi}{\partial \overset{4}{\varepsilon}} \tag{3.14}$$

When the terms $\overset{4}{\boldsymbol{T}}$ and $\overset{4}{\varepsilon}$ are dropped, the first strain gradient theory is recovered. Korteweg's stress $\boldsymbol{\tau}$ can then be interpreted as the effective stress of the first gradient of strain theory, as done in Casal and Gouin (1985).

Let us specify the constitutive equations in the case of an elastic fluid. Mindlin considers that, in an elastic fluid, all stress tensors are spherical so that

$$\overset{2}{\boldsymbol{T}} = -p\boldsymbol{1}, \quad \overset{3}{\boldsymbol{T}} = -\boldsymbol{1} \otimes \overset{1}{\boldsymbol{p}}, \quad \overset{4}{\boldsymbol{T}} = -\boldsymbol{1} \otimes \overset{2}{\boldsymbol{\pi}}, \tag{3.15}$$

where Mindlin's notations for the introduced zeroth, first and second-order stress tensors p, $\overset{1}{\boldsymbol{p}}$ and $\overset{2}{\boldsymbol{\pi}}$ have been kept. As a result, the effective stress is spherical and reads

$$\boldsymbol{\tau} = -(p - \overset{1}{\boldsymbol{p}} \cdot \nabla + (\overset{2}{\boldsymbol{\pi}} : (\nabla \otimes \nabla))\,\boldsymbol{1}, \tag{3.16}$$

the divergence of which vanishes. It is in contrast to Korteweg's medium, see Eq. (3.10), which can transmit shear stresses. The concept of an elastic fluid has been clearly defined by Noll for simple materials (Truesdell and Noll 1965) and leads to a spherical Cauchy stress tensor. The question of the definition of a first strain gradient elastic fluid has been only tackled by Fried and Gurtin (2006). Clearly, Korteweg's and Mindlin's definitions of an elastic first strain gradient fluid differ. Mindlin's formulation is less general than that given by Eq. (3.10). Following Mindlin's definition, the free energy potential of an isotropic second strain gradient linear elastic fluid takes the form

$$\rho\Psi(\Delta, \nabla\Delta, \nabla\otimes\nabla\Delta) = \frac{\lambda}{2}\Delta^2 + a_1(\nabla\Delta)^2 + b_0\nabla^2\Delta + b_1(\nabla^2\Delta)^2 + b_2(\nabla\otimes\nabla\Delta) : (\nabla\otimes\nabla\Delta) + c_1\Delta\nabla^2\Delta \quad (3.17)$$

All terms are quadratic except the linear term that involves the material parameter b_0. The stress–strain relations follow

$$p = -\rho\frac{\partial\Psi}{\partial\Delta} = -\lambda\Delta - c_1\nabla^2\Delta, \quad \overset{1}{p} = -\rho\frac{\partial\Psi}{\partial\nabla\Delta} = -2a_1\nabla\Delta, \quad (3.18)$$

$$\overset{2}{\pi} = -\rho\frac{\partial\Psi}{\partial(\nabla\otimes\nabla\Delta)} = -(b_0 + c_1\Delta + 2b_1\nabla^2\Delta)\mathbf{1} - 2b_2\nabla\otimes\nabla\Delta. \quad (3.19)$$

It contains, in general, a self-equilibrated higher-order stress $\overset{2}{\pi} = -b_0\mathbf{1}$.

The effective stress then becomes

$$\tau = -(-\lambda\Delta + 2(a_1 - c_1)\nabla^2\Delta - 2(b_1 + b_2)\nabla^4\Delta)\mathbf{1} \quad (3.20)$$

$$\tau_{ij} = -(-\lambda\Delta + 2(a_1 - c_1)\Delta_{,kk} - 2(b_1 + b_2)\Delta_{,kkll})\delta_{ij}. \quad (3.21)$$

The corresponding effective stress for the first strain gradient model is obtained by setting $b_1 = b_2 = 0$. It can be directly compared to Korteweg's equation (3.10). Both relations contain the Laplacian term, $\nabla^2\Delta$. Korteweg's equation contains an additional non-spherical term, as already mentioned. Mindlin's second gradient of strain theory introduces an additional contribution of fourth order.

Combining the balance and constitutive equations, the following fourth-order partial differential equation is derived

$$(1 - l_1^2\nabla^2)(1 - l_2^2\nabla^2)\Delta = 0, \quad (3.22)$$

that involves two characteristic lengths such that

$$\lambda l_i^2 = a_1 - c_1 \pm \sqrt{(a_1 - c_1)^2 - 2\lambda(b_1 + b_2)}, \quad i = 1, 2. \quad (3.23)$$

Note that Eq. (3.22) follows from a first integration of the fifth-order balance equation (3.13). Accordingly, a constant term should be added in the right-hand side of the equation, leading to a homogeneous pressure field to be superimposed to the solution. The Eq. (3.22) can be referred to as a bi-Helmholtz equation.

Let us now consider capillary effects in a spherical homogeneous material made of a second strain gradient isotropic linear elastic fluid. We look for a dilatation field

$\Delta(r)$, and equations are solved in spherical coordinates. In that specific case, Mindlin shows that the dilatation field is then of the form

$$\Delta = \frac{C_1 l_1}{r_1} \sinh \frac{r}{l_1} + \frac{C_2 l_2}{r_2} \sinh \frac{r}{l_2}. \tag{3.24}$$

The integration constants C_1 and C_2 must be determined from the boundary conditions. A key point of the analysis is that the outer surface of the droplet of initial radius r_0 is assumed to be free of traction forces. Mindlin shows that these traction-free boundary conditions lead to two equations for the unknowns C_1 and C_2:

$$\sum_{i=1}^{2} C_i r_i \left(2(b_1 + b_2) - (2a_1 - c_1) l_i^2\right) (r_i \cosh r_i - \sinh r_i) = 0, \tag{3.25}$$

$$\sum_{i=1}^{2} C_i \left(r_i \left(2b_2 (1 + \frac{2}{r_i}) + 2b_1 + c_1 l_i^2 \right) \sinh r_i - 4b_2 \cosh r_i \right) = -b_0 r_0^2 \tag{3.26}$$

where $r_i = r_0 / l_i$. When the radius of curvature is large in comparison with the characteristic lengths l_i (i.e., $r_i \gg 1$), Mindlin derives a relation between the mean dilatation of the droplet and its radius of curvature

$$\bar{\Delta} = \frac{3}{4\pi r_0^3} \int_V \Delta \, dV = -\frac{6c_1}{b_0 \lambda} \frac{\gamma}{r_0}, \tag{3.27}$$

where γ can be identified with the surface tension in the Laplace–Young equation (3.1) and takes the value

$$\gamma = \frac{T_0}{1 + (8b_2 T_0 / b_0^2 r_0)}, \quad \text{with} \quad T_0 = \frac{b_0 (l_1^2 - l_2^2)}{2\lambda (l_1 (l_2^2 + c_1^2/\lambda^2)^2 - l_2 (l_1^2 + c_1^2/\lambda^2)^2)}. \tag{3.28}$$

The identification between the diffuse and sharp interface models for surface tension shows that the relevant material parameters that intervene in the relation are c_1, b_2 and b_0 which are all related to the contributions of the second strain gradient in the elastic potential (3.17). If $b_0 = 0$, the system of equations (3.25)–(3.26) is homogeneous and leads to the trivial solution $C_1 = C_2 = 0$ and to a homogeneous mass density field. The internal hyperstresses account for material cohesion which collapses at a free surface.

The corresponding balance equations and solutions of the spherical droplet can also be obtained for the first gradient of strain theory, in the same way. For, the first strain gradient theory can be formally obtained as a special case of the second strain gradient model by setting that the fourth-order stress tensor $\overset{4}{\boldsymbol{T}}$ vanishes in Eq. (3.11), or, equivalently, that the second characteristic length l_2 vanishes in (3.22). The Eq. (3.22) then reduces to

$$(1 - l^2\nabla^2)\Delta = 0, \quad \text{with} \quad l^2 = \frac{2a_1}{\lambda}, \tag{3.29}$$

which involves only one characteristic length l. There is still a solution of the form

$$\Delta = \frac{C}{r} \sinh \frac{r}{l_1} \tag{3.30}$$

for the problem of the spherical droplet made of a linear elastic isotropic medium. However, the condition of vanishing simple and double tractions at the free outer surface leads to a homogeneous equation so that the integration constant is identified as $C = 0$. This leads to the trivial solution $\Delta = 0$ and homogeneous mass density within the droplet.

Accordingly, the description of capillarity effects in a linear elastic medium requires the introduction of the second derivatives of the strain in the free energy density. The situation is different for a liquid droplet in equilibrium with vapor that can be described by a Cahn–Hilliard fluid model, also a higher-order gradient theory (Seppecher 2000).

Second Strain Gradient Elasticity of Nano-objects

Strain gradient elasticity is now applied to isotropic elastic solids following initial ideas by Mindlin (1965) and recent experiment results on the tension of nano-wires. The links between surface energy and surface elasticity are discussed for elastic second strain gradient materials. The presented results are reported from Cordero et al. (2016).

Balance Equations and Boundary Conditions

Mindlin's third gradient material is an elastic solid endowed with an Helmholtz free energy density function that depends on the strain, strain gradient and second gradient of the strain tensors (Mindlin 1965):

$$\rho\Psi = \rho\Psi\left(\varepsilon,\ \varepsilon \otimes \nabla,\ \varepsilon \otimes \nabla \otimes \nabla\right), \tag{3.31}$$

The theory can also be formulated in terms of the first, second and third gradients of the displacement field $\boldsymbol{u}$:

$$\rho\Psi = \rho\tilde{\Psi}\left(\varepsilon,\ \overset{3}{\varepsilon} = \boldsymbol{u} \otimes \nabla \otimes \nabla,\ \overset{4}{\varepsilon} = \boldsymbol{u} \otimes \nabla \otimes \nabla \otimes \nabla\right), \tag{3.32}$$

Both formulations are equivalent due to compatibility requirements that imply bijective relationships between the strain gradient and the second gradient of the displacement field:

$$\varepsilon_{ij,k} = \frac{1}{2}\left(u_{i,jk} + u_{j,ik}\right), \quad u_{i,jk} = \varepsilon_{ij,k} + \varepsilon_{ki,j} - \varepsilon_{jk,i}. \tag{3.33a}$$

Next, the theory is exploited using the expression (3.32) in terms of strain and second and third gradients of the displacement field. The classical infinitesimal strain tensor, ε, is the symmetric part of the displacement field gradient and has six independent components in 3D. The second gradient of the displacement field, $\overset{3}{\varepsilon}$, is symmetric in the last two indices and has eighteen independent components while the third gradient of the displacement field, $\overset{4}{\varepsilon}$, is symmetric in the last three indices and has thirty independent components. In 2D, these tensors, respectively, have four, six and eight independent components. The conjugate stress tensors $\overset{2}{\boldsymbol{T}}, \overset{3}{\boldsymbol{T}}, \overset{4}{\boldsymbol{T}}$ are obtained from the state laws given by Eq. (3.14). The generalized stress tensors share the same symmetry properties as ε, $\overset{3}{\varepsilon}$ and $\overset{4}{\varepsilon}$, respectively. They must fulfill the balance law (3.13).

The work of contact forces takes the form

$$w^{(ext)} = \boldsymbol{t}^1 \cdot \boldsymbol{u} + \boldsymbol{t}^2 \cdot D_n \boldsymbol{u} + \boldsymbol{t}^3 \cdot D_{nn} \boldsymbol{u} \tag{3.34}$$

where $\boldsymbol{t}^1$, $\boldsymbol{t}^2$ and $\boldsymbol{t}^3$ are generalized surface traction vectors and D_n, D_{nn} are first- and second-order normal derivatives of the displacement field. The gradient of $\boldsymbol{u}$ on a surface ∂V is decomposed into a normal gradient and a surface gradient:

$$\boldsymbol{u} \otimes \nabla = (D_n \boldsymbol{u}) \otimes \boldsymbol{n} + \boldsymbol{u} \otimes \boldsymbol{D}_t, \quad u_{i,j} = D_n u_i \, n_j + D_{tj} u_i \tag{3.35}$$

where $\boldsymbol{n}$ is the unit normal vector at a surface point and

$$D_n \boldsymbol{u} := (\boldsymbol{u} \otimes \nabla) \cdot \boldsymbol{n}, \quad D_n u_i := u_{i,k} \, n_k \tag{3.36}$$

$$D_{nn} \boldsymbol{u} := (\boldsymbol{u} \otimes \nabla \otimes \nabla) : (\boldsymbol{n} \otimes \boldsymbol{n}), \quad D_{nn} u_i := u_{i,kl} \, n_k \, n_l \tag{3.37}$$

The surface gradient operator, $\boldsymbol{D}_t$, follows then from Eq. (3.35).

The traction boundary conditions associated with the balance law (3.13) read

$$\begin{aligned} \boldsymbol{t}^1 = \Big[\overset{2}{\boldsymbol{T}} - \overset{3}{\boldsymbol{T}} \cdot \nabla + \overset{4}{\boldsymbol{T}} : (\nabla \otimes \nabla) \Big] . \boldsymbol{n} + \Big[\Big(\overset{3}{\boldsymbol{T}} - \overset{4}{\boldsymbol{T}} . \nabla \Big) . \boldsymbol{n} + \overset{4}{\boldsymbol{T}} : (\boldsymbol{n} \otimes \boldsymbol{L}) \\ - \Big(\overset{4}{\boldsymbol{T}} : (\boldsymbol{n} \otimes \boldsymbol{n}) \Big) . (\boldsymbol{n} \otimes \boldsymbol{D}_t) \Big] . \boldsymbol{L} \end{aligned} \tag{3.38a}$$

$$\boldsymbol{t}^2 = \Big(\overset{3}{\boldsymbol{T}} - \overset{4}{\boldsymbol{T}} . \nabla \Big) : (\boldsymbol{n} \otimes \boldsymbol{n}) + \overset{4}{\boldsymbol{T}} \vdots (\boldsymbol{n} \otimes \boldsymbol{L} \otimes \boldsymbol{n}) + \overset{4}{\boldsymbol{T}} \vdots (\boldsymbol{n} \otimes \boldsymbol{n} \otimes \boldsymbol{L}) \tag{3.38b}$$

$$t^3 = \overset{4}{T} \vdots (n \otimes n \otimes n)\,. \tag{3.38c}$$

Mindlin's linear surface operator $\boldsymbol{L}$ is expressed as a function of the surface gradient operator, $\boldsymbol{D}_t$, by its action on a vector field $\boldsymbol{\phi}$:

$$\boldsymbol{L} \cdot \boldsymbol{\phi} = (\boldsymbol{D}_t \cdot \boldsymbol{n})\, \boldsymbol{\phi} \cdot \boldsymbol{n} - \boldsymbol{D}_t \cdot \boldsymbol{\phi} \tag{3.39}$$

The corresponding Dirichlet conditions consist in prescribing the displacement vector and its first and second normal derivatives at the boundary.

Constitutive Equations in Isotropic Linear Elasticity

Mindlin (1965) derived the strain energy density for isotropic linear elastic second strain gradient materials as follows:

$$\begin{aligned} \rho\Psi(\varepsilon_{ij}, \varepsilon_{ijk}, \varepsilon_{ijkl}) = & \frac{1}{2}\lambda\, \varepsilon_{ii}\, \varepsilon_{jj} + \mu\, \varepsilon_{ij}\, \varepsilon_{ij} + a_1\, \varepsilon_{iij}\, \varepsilon_{kkj} + a_2\, \varepsilon_{ijj}\, \varepsilon_{kki} \\ & + a_3\, \varepsilon_{ijj}\, \varepsilon_{ikk} + a_4\, \varepsilon_{ijk}\, \varepsilon_{ijk} + a_5\, \varepsilon_{ijk}\, \varepsilon_{jik} \\ & + b_1\, \varepsilon_{iijj}\, \varepsilon_{kkll} + b_2\, \varepsilon_{iikl}\, \varepsilon_{jjkl} + b_3\, \varepsilon_{ijkk}\, \varepsilon_{llji} \\ & + b_4\, \varepsilon_{ijkk}\, \varepsilon_{jill} + b_5\, \varepsilon_{ijkk}\, \varepsilon_{ijll} + b_6\, \varepsilon_{ijkl}\, \varepsilon_{ijkl} + b_7\, \varepsilon_{ijkl}\, \varepsilon_{jkli} \\ & + c_1\, \varepsilon_{ii}\, \varepsilon_{jjkk} + c_2\, \varepsilon_{ij}\, \varepsilon_{kkij} + c_3\, \varepsilon_{ij}\, \varepsilon_{jikk} + b_0\, \varepsilon_{iijj}\,. \end{aligned} \tag{3.40}$$

Sixteen elasticity moduli arise in the free energy function in addition to the usual Lamé constants, λ and μ. The following distinction can be made among the numerous terms in Eq. (3.40):

- The five parameters, a_i, are the higher-order elasticity moduli related to the second gradient of displacement or, equivalently, first strain gradient part of the theory, as derived by Toupin (1962), Mindlin (1964);
- The seven parameters, $b_{i,\, i\neq 0}$, are the higher-order elasticity moduli specifically related to the second strain gradient;
- The three parameters, c_i, are coupling moduli responsible for the coupling between the strain and the third gradient of the displacement field;
- The initial higher-order stress or cohesion modulus, b_0, is related to the surface energy.

It will be shown that the surface elasticity effects are accounted for by the coupling moduli, c_i all appearing in the third-order terms of the theory. The constitutive equations are obtained from the free energy density (3.40) and the definitions (3.14). They read

$$T_{pq} = \lambda\, \varepsilon_{ii}\, \delta_{pq} + 2\, \mu\, \varepsilon_{pq} + c_1\, \varepsilon_{iijj}\, \delta_{pq} + c_2\, \varepsilon_{iipq} + \frac{1}{2}\, c_3\, \left(\varepsilon_{pqii} + \varepsilon_{qpii}\right), \quad (3.41a)$$

$$T_{pqr} = a_1\, \left(\varepsilon_{iir}\, \delta_{pq} + \varepsilon_{iiq}\, \delta_{pr}\right) + \frac{1}{2}\, a_2\, \left(\varepsilon_{rii}\, \delta_{pq} + 2\, \varepsilon_{iip}\, \delta_{qr} + \varepsilon_{qii}\, \delta_{pr}\right)$$
$$+\, 2\, a_3\, \varepsilon_{pii}\, \delta_{qr} + 2\, a_4\, \varepsilon_{pqr} + a_5\, \left(\varepsilon_{qrp} + \varepsilon_{rqp}\right), \quad (3.41b)$$

$$T_{pqrs} = \frac{2}{3}\, b_1\, \varepsilon_{iijj}\, \delta_{pqrs} + \frac{2}{3}\, b_2\, \varepsilon_{iijk}\, \delta_{jkpqrs}$$
$$+ \frac{1}{6}\, b_3\, \left(\left(\varepsilon_{jkii} + \varepsilon_{kjii}\right) \delta_{jkpqrs} + 2\varepsilon_{iijp}\, \delta_{jqrs}\right) + \frac{2}{3}\, b_4\, \varepsilon_{jpii}\, \delta_{jqrs}$$
$$+ \frac{2}{3}\, b_5\, \varepsilon_{pjii}\, \delta_{jqrs} + 2\, b_6\, \varepsilon_{pqrs} + \frac{2}{3}\, b_7\, \left(\varepsilon_{qrsp} + \varepsilon_{rspq} + \varepsilon_{spqr}\right)$$
$$+ \frac{1}{3}\, c_1\, \varepsilon_{ii}\, \delta_{pqrs} + \frac{1}{3}\, c_2\, \varepsilon_{ij}\, \delta_{ijpqrs} + \frac{1}{3}\, c_3\, \varepsilon_{ip}\, \delta_{iqrs} + \frac{1}{3}\, b_0\, \delta_{pqrs}, \quad (3.41c)$$

where fourth- and sixth-order identity tensors are defined as follows:

$$\delta_{ijkl} = \delta_{ij}\delta_{kl} + \delta_{ik}\delta_{jl} + \delta_{jk}\delta_{il}, \quad \delta_{ijklmn} = \delta_{im}\delta_{jn}\delta_{kl} + \delta_{il}\delta_{jn}\delta_{km} + \delta_{il}\delta_{jm}\delta_{kn} \quad (3.42)$$

with δ_{ij} being the Kronecker symbol.

Surface Energy Effects in Third Gradient Elasticity

In this section, surface energy-related effects are evidenced in one simple boundary value problem involving a free surface. The detailed analytic solutions also provide expressions of several characteristic lengths defined as functions of the isotropic elasticity moduli (3.40). Corresponding stability requirements are discussed in Cordero et al. (2016), Khakalo and Niiranen (2018). These are essentially one-dimensional solutions which involve several intrinsic lengths and dimensionless parameters that will be used in section "Surface Elasticity Effect."

Half-Space with a Free Surface

The half-space, $x_1 \geq 0$, in a Cartesian coordinate system $\{x_1, x_2, x_3\}$ is occupied by an elastic second strain gradient material. In continuum mechanics, free surfaces are associated with Neumann conditions of vanishing tractions. In the context of a second strain gradient theory, the planar surface, $x_1 = 0$, is traction-free if

$$\boldsymbol{t}^1 = \boldsymbol{t}^2 = \boldsymbol{r}^3 = 0 \quad \text{at} \quad x_1 = 0. \quad (3.43)$$

We look for displacement fields of the form

$$u_1 = u_1(x_1), \quad u_2 = u_3 = 0. \tag{3.44}$$

The current problem is then essentially one dimensional. The stress equation of equilibrium (3.13) becomes

$$T_{11,1} - T_{111,11} + T_{1111,111} = 0 \tag{3.45}$$

in the absence of body forces. The boundary conditions (3.38) then write

$$T_{11} - T_{111,1} + T_{1111,11} = 0, \quad T_{111} - T_{1111,1} = 0, \quad T_{1111} = 0, \quad \text{at} \quad x_1 = 0 \tag{3.46}$$

With the assumed displacement field Eq. (3.44), the potential energy density (3.40) is a function of the strain components ε_{11}, ε_{111} and ε_{1111} only. Thus,

$$\rho\Psi\left(\varepsilon_{11}, \varepsilon_{111}, \varepsilon_{1111}\right) = \left(\frac{\lambda}{2} + \mu\right)\varepsilon_{11}^2 + \frac{A}{2}\varepsilon_{111}^2 + \frac{B}{2}\varepsilon_{1111}^2 + \bar{c}\,\varepsilon_{11}\,\varepsilon_{1111} + b_0\,\varepsilon_{1111} \tag{3.47}$$

with the following notations for reduced moduli,

$$A = 2\left(a_1 + a_2 + a_3 + a_4 + a_5\right), \tag{3.48}$$

$$B = 2\left(b_1 + b_2 + b_3 + b_4 + b_5 + b_6 + b_7\right), \tag{3.49}$$

$$\bar{c} = c_1 + c_2 + c_3. \tag{3.50}$$

The constants A and $\bar{c}$ have the dimension of forces (N) and B has the dimension of force times unit surface (N.mm^2). The material parameters L_1 and L_2 with the dimension of length are now defined, together with the dimensionless parameters η and η_0:

$$L_1^2 = \frac{A}{\lambda/2 + \mu}, \quad L_2^2 = \frac{B}{A}, \quad \eta = \frac{\bar{c}}{A}, \quad \eta_0 = \frac{b_0}{A}. \tag{3.51}$$

The next step here is to rewrite the potential energy density, Eq. (3.47), as a function of three-dimensionless arguments:

$$\rho\Psi\left(\varepsilon_{11}, \varepsilon_{111}, \varepsilon_{1111}\right) = \rho\hat{\Psi}\left(\varepsilon_{11}, L_1\,\varepsilon_{111}, L_2^2\,\varepsilon_{1111}\right) \tag{3.52}$$

Stability requirements, i.e., convexity of the free energy potential implies that the moduli A and B are positive. However, the moduli $\bar{c}$ and b_0 can be either positive or negative.

Combining the Eqs. (3.44) and (3.41), the components T_{11}, T_{111} and T_{1111} of the stress tensors are derived as follows:

$$T_{11} = (\lambda + 2\mu)\,\varepsilon_{11} + \bar{c}\,\varepsilon_{1111}, \tag{3.53a}$$

$$T_{111} = A\,\varepsilon_{111}, \tag{3.53b}$$

$$T_{1111} = B\,\varepsilon_{1111} + \bar{c}\,\varepsilon_{11} + b_0. \tag{3.53c}$$

Substituting these expressions into the stress equation of equilibrium (3.45) leads to the following sixth-order displacement equation of equilibrium:

$$\left(1 - l_1^2 \frac{d^2}{dx_1^2}\right)\left(1 - l_2^2 \frac{d^2}{dx_1^2}\right)\frac{d^2 u_1}{dx_1^2} = 0. \tag{3.54}$$

It involves new characteristic lengths defined in the same way as in Mindlin (1965), namely

$$l_1^2 l_2^2 = \frac{B}{\lambda + 2\mu}, \quad l_1^2 + l_2^2 = \frac{A - 2\bar{c}}{\lambda + 2\mu}, \tag{3.55}$$

from which the individual intrinsic lengths, l_1 and l_2, can be obtained:

$$l_1^2 = \frac{A - 2\bar{c} + \sqrt{(A - 2\bar{c})^2 - 4B\,(\lambda + 2\mu)}}{2\,(\lambda + 2\mu)},$$

$$l_2^2 = \frac{A - 2\bar{c} - \sqrt{(A - 2\bar{c})^2 - 4B\,(\lambda + 2\mu)}}{2\,(\lambda + 2\mu)}$$

or alternatively:

$$l_1^2 = \frac{L_1^2}{4}\left(1 - 2\eta + \sqrt{(1 - 2\eta)^2 - 8\frac{L_2^2}{L_1^2}}\right),$$

$$l_2^2 = \frac{L_1^2}{4}\left(1 - 2\eta - \sqrt{(1 - 2\eta)^2 - 8\frac{L_2^2}{L_1^2}}\right),$$

using the previously defined characteristic lengths L_1 and L_2. The solution of (3.54) then has the general form:

$$u_1(x_1) = \alpha_1\, e^{x_1/l_1} + \alpha_2\, e^{-x_1/l_1} + \alpha_3\, e^{x_1/l_2} + \alpha_4\, e^{-x_1/l_2} + \alpha_5\, x_1 + \alpha_6. \tag{3.56}$$

Three different types of material responses can be described by these solutions: exponential decaying, aperiodic and harmonic depending on the characteristic lengths l_1 and l_2 considered as complex numbers; see the discussion in Cordero et al. (2016). It can be proved that no surface effect can occur in the absence of initial higher-order stress, i.e., when $b_0 = 0$.

The corresponding displacement and stress profiles close to the traction-free surface are presented in Fig. 3.1. These profiles are obtained for $L_1 = 0.3$ nm, a fixed ratio $L_2^2/L_1^2 = 1$, $\eta_0 = 0.1$ and for four different values of η. These values are chosen to illustrate the three regimes. Note that the harmonic solution which does not vanish at infinity is unstable. The profiles presented in Fig. 3.1 show that the surface energy effects produced by the theory can either lead to higher or smaller inter-atomic distances close to the free surface. The free surface and subsurface material can therefore be either in tension or compression. It is remarkable that, within the second strain

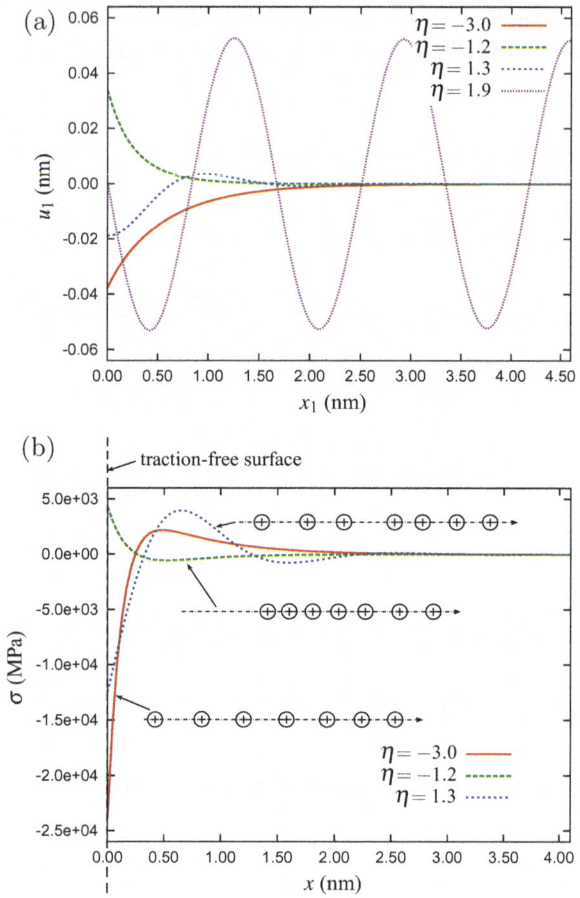

Fig. 3.1 Profiles of **a** the displacement component u_1, and **b** the stress component $T = T_{11}$ close to the traction-free surface (at $x_1 = 0$). Four sets of material parameters are considered as follows: $E = 140$ GPa, $\nu = 0.3$, $L_1 = 0.3$ nm and a fixed ratio $L_2^2/L_1^2 = 1$. The surface energy related parameter $\eta_0 = 0.1$ and four values of η are used in order to illustrate various cases arising in the theory. After Cordero et al. (2016). The circles along a line symbolically denote the position of atoms close to the free surface to enlighten the variation of spacing between them

gradient theory with an initial cohesion higher-order stress, the traction-free condition does not lead to a vanishing T_{11}, as could be expected from Eq. $(3.46)_1$.

Surface Elasticity Effect

Casal (1961, 1963) was the first to theoretically consider the tension of a nano-wire made of an isotropic first strain gradient material. He derived the boundary layers effects at the ends of a beam induced by higher-order boundary conditions. However, he did not mention the non-homogeneous straining close to a free surface. This is due to the fact that such effects cannot arise in isotropic first strain gradient materials, as discovered by Mindlin (1965). These phenomena akin to surface elasticity are now illustrated for the second strain gradient continuum.

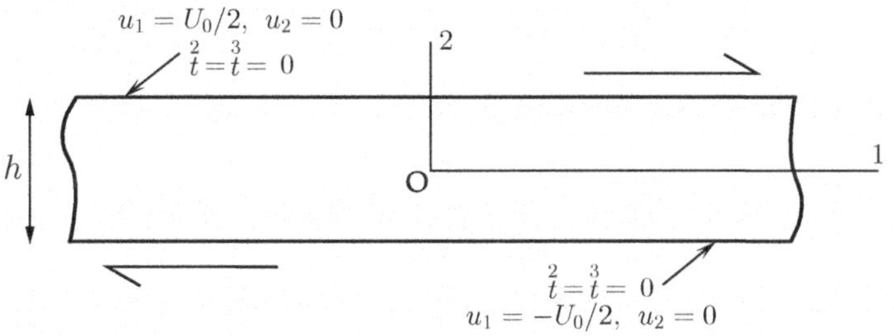

Fig. 3.2 Shear of an infinite strip of thickness h. Dirichlet displacement conditions are applied to the upper and lower surfaces. These surfaces are free of higher-order tractions

The purpose of this section is to show how the theory can describe surface effects induced by the coupling between the strain, ε, and the third gradient of the displacement field, $\overset{4}{\varepsilon}$. This coupling is controlled by the moduli c_i, as can be seen from the constitutive equations (3.41a) and (3.41c).

The tension of nano-wires/nano-films and their apparent elastic properties are investigated based on finite element simulations to be presented in section "Apparent Tensile Elastic Behavior of Second Strain Gradient Nano-Films." The notion of apparent elastic property is first defined analytically in the context of third gradient elasticity. Analytic expressions for the apparent shear modulus are derived for an infinite thin strip subjected to shear.

Resolution of the Boundary Value Problem

The simple shear of an infinite strip of thickness h is considered as described in Fig. 3.2. Displacements are prescribed to the upper and lower surfaces corresponding to simple shear:

$$u_1(h/2) = \frac{U_0}{2}, \quad u_1(-h/2) = -\frac{U_0}{2}. \tag{3.57}$$

These upper and lower surfaces are assumed to be free of higher-order tractions, i.e.,

$$\overset{2}{t} = \overset{3}{t} = 0 \quad \text{at} \quad x_2 = \pm\frac{h}{2}, \tag{3.58}$$

In addition, in order to focus exclusively on the surface elasticity effects, no surface energy is considered, which amounts to $b_0 = 0$. Due to the infinite lateral extension of the strip, the displacement field takes the form:

$$u_1 = u_1(x_2), \quad u_2 = u_3 = 0. \tag{3.59}$$

The stress balance equation (3.13) becomes

$$T_{12,2} - T_{122,22} + T_{1222,222} = 0 \tag{3.60}$$

and the two last boundary conditions (3.38) are

$$T_{122} - T_{1222,2} = T_{1222} = 0 \quad \text{at} \quad x_2 = \pm\frac{h}{2}. \tag{3.61}$$

The strain energy density (3.40) is a function of the strain components ε_{12}, ε_{122} and ε_{1222}:

$$\rho\Psi(\varepsilon_{12}, \varepsilon_{122}, \varepsilon_{1222}) = \mu\,\varepsilon_{12}^2 + \frac{A}{2}\,\varepsilon_{122}^2 + \frac{B}{2}\,\varepsilon_{1222}^2 + c_3\,\varepsilon_{12}\,\varepsilon_{1222} \tag{3.62}$$

with the new definitions of reduced moduli A and B superseding (3.50),

$$A = 2\,(a_3 + a_4)\,, \quad B = 2\,(b_5 + b_6)\,. \tag{3.63}$$

In the same way as in section "Half-Space with a Free Surface," the following reduced material parameters are introduced

$$L_1^2 = \frac{A}{\mu}, \quad L_2^2 = \frac{B}{A}, \quad \eta = \frac{c_3}{A}, \tag{3.64}$$

which are used to express the potential energy density (3.62) as a function of dimensionless arguments:

$$\rho\Psi(\varepsilon_{12}, \varepsilon_{122}, \varepsilon_{1222}) = \rho\hat{\Psi}\left(\varepsilon_{12}, L_1\,\varepsilon_{122}, L_2^2\,\varepsilon_{1222}\right). \tag{3.65}$$

The following requirements can then be obtained based on convexity of the strain energy potential,

$$L_1^2 \geq 0, \quad L_2^2 \geq 0, \quad 2\,\frac{L_2^2}{L_1^2} \geq \eta^2. \tag{3.66}$$

This implies that the moduli A and B must be positive while c_3 can be either positive or negative. The components T_{12}, T_{122} and T_{1222} of the stress tensors are obtained from Eqs. (3.62) and (3.41):

$$T_{12} = 2\mu\,\varepsilon_{12} + c_3\,\varepsilon_{1222}, \tag{3.67a}$$

$$T_{122} = A\,\varepsilon_{122}, \tag{3.67b}$$

$$T_{1222} = B\,\varepsilon_{1222} + c_3\,\varepsilon_{12}. \tag{3.67c}$$

Substituting these expressions into the stress equation of equilibrium (3.60) and recalling that $\varepsilon_{12} = 1/2\,u_{1,2}$, $\varepsilon_{122} = 1/2\,u_{1,22}$ and $\varepsilon_{1222} = 1/2\,u_{1,222}$, the following displacement equation for equilibrium is obtained:

$$\left(1 - l_1^2 \frac{d^2}{dx_2^2}\right)\left(1 - l_2^2 \frac{d^2}{dx_2^2}\right)\frac{d^2 u_1}{dx_2^2} = 0. \tag{3.68}$$

The relationships

$$l_1^2 l_2^2 = \frac{B}{2\mu}, \quad l_1^2 + l_2^2 = \frac{A - 2c_3}{2\mu}, \tag{3.69}$$

are obtained and must be used to derive expressions of the lengths l_1 and l_2.

The solution of Eq. (3.68) takes the form

$$u_1(x_2) = \alpha_1\, e^{x_2/l_1} + \alpha_2\, e^{-x_2/l_1} + \alpha_3\, e^{x_2/l_2} + \alpha_4\, e^{-x_2/l_2} + \alpha_5\, x_2 + \alpha_6, \tag{3.70}$$

and the boundary conditions (3.61) require that

$$(A - c_3)\,\varepsilon_{122} - B\,\varepsilon_{1222,2} = 0, \quad B\,\varepsilon_{1222} + c_3\,\varepsilon_{12} = 0 \qquad \text{at}\, x_2 = \pm\frac{h}{2}, \tag{3.71}$$

They are used to find the expressions of the constants α_1 to α_6 that can be found in Cordero et al. (2016). It can be shown from these expressions that surface effects only occur if a coupling exists between the component of the strain ε_{12} and the component of the third gradient of the displacement field ε_{1222}. If the coupling modulus $c_3 = 0$, then $\alpha_1 = \alpha_2 = \alpha_3 = \alpha_4 = 0$ and $\alpha_5 = U_0/h$. The standard homogeneous shear solution is retrieved: $u_1(x_2) = U_0 x_2/h$.

Determination of the Apparent Shear Modulus

Experimental measurements of apparent elastic properties of nano-objects are based on the ratio of effective stress and strain properties, since the direct measurement of higher-order stresses remains an open question. Such apparent quantities can be derived from the previous theory by proper averaging procedures.

The apparent shear modulus μ^{app} is defined by averaging the stress component, T_{12}, and the strain component, ε_{12}, along the thickness of the infinite strip as follows:

$$\int_{-h/2}^{h/2} T_{12}\, dx_2 = 2\,\mu^{app} \int_{-h/2}^{h/2} \varepsilon_{12}\, dx_2. \tag{3.72}$$

The following expression of the apparent shear modulus is then obtained from the analytic solution:

$$\frac{\mu}{\mu^{app}} = 1 - \frac{2}{h}\,\frac{l_c^4\,(l_1^2 - l_2^2)}{l_1\,(l_2^2 + l_c^2)^2 \coth\left(\dfrac{h}{2l_1}\right) - l_2\,(l_1^2 + l_c^2)^2 \coth\left(\dfrac{h}{2l_2}\right)}. \tag{3.73}$$

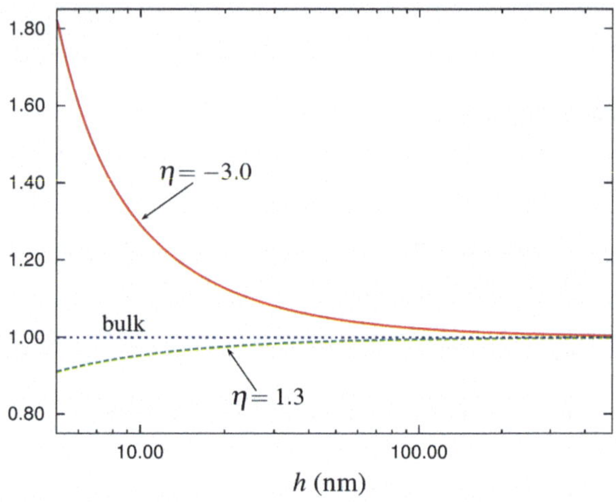

Fig. 3.3 Effect of the thickness of the infinite strip, h, on the evolution of the ratio μ^{app}/mu, for $E = 140$ GPa, $\nu = 0.3$, $L_1 = 0.3$ nm, $L_2^2/L_1^2 = 1$ and for two different values of the coupling parameter, η. The horizontal line corresponds to the classical solution obtained without coupling ($\eta = 0$), or equivalently, without surface stress effects. Note that in all of these cases, the surface energy is not considered, i.e., $b_0 = \eta_0 = 0$

It is apparent in this relation that μ^{app} is size-dependent, the effect of the thickness h on its value is plotted for different sets of material parameters in Fig. 3.3. For large values of h, the surface elasticity effects are negligible and the bulk shear modulus μ is retrieved,

$$\lim_{h\to\infty} \mu^{app} = \mu. \tag{3.74}$$

When the strip thickness decreases, the apparent elastic property progressively departs from the one of the bulk and tends to the limit

$$\lim_{h\to 0} \mu^{app} = \mu \left(1 - \frac{l_c^4}{l_1^2 l_2^2}\right). \tag{3.75}$$

Note that when the coupling modulus c_3 vanishes, the ratio μ^{app}/μ is equal to one and no surface elasticity effect arises. Figure 3.3 shows that the apparent shear modulus, μ^{app}, can either increase or decrease for smaller h. This only depends on the material parameter related to the coupling, namely η controlling the sign of l_1^2 and l_2^2. The same dependency appears in the limit presented in Eq. (3.75).

Apparent Tensile Elastic Behavior of Second Strain Gradient Nano-Films

The size-dependent tensile behavior of nano-objects, especially nano-wires, has been widely investigated due to their key role in nano-electromechanical systems (NEMS) (Thomas et al. 2011). In particular, their size-dependent elastic properties are studied experimentally using various methods (see, for example, Agrawal et al. 2008;

McDowell et al. 2008; Sadeghian et al. 2009). Nano-wires exhibit a very high surface area to volume ratio, and therefore, their behavior is strongly affected by surface effects. Using the previous results as a guideline to define the material parameters, the overall behavior of such nano-objects and especially the local strain fields are studied using finite element simulations described in more detail in Cordero et al. (2016).

The boundary value problem solved by means of the finite element method is depicted in Fig. 3.4. Here, a long plate with thickness, d, is subjected to prescribed displacement $u_2 = U$ at the top in the direction 2 and fixed displacement u_2 at the bottom. Plane strain conditions are enforced, i.e., $u_3 = 0$. Periodicity conditions are

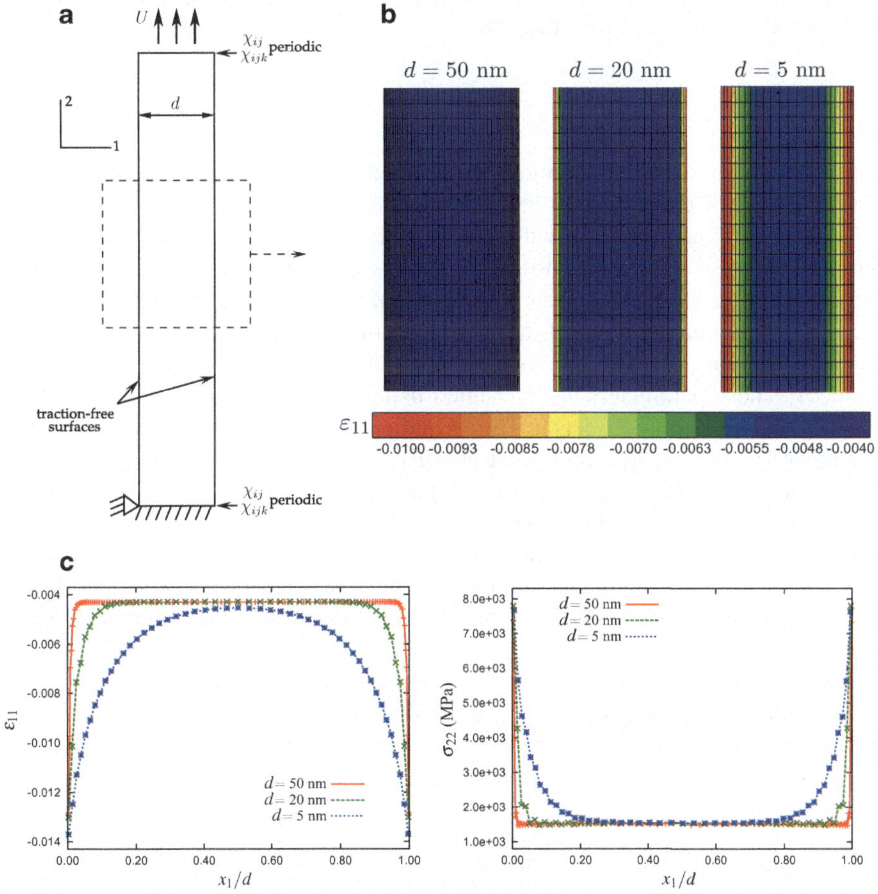

Fig. 3.4 Finite element simulations of tensile tests of thin films with material parameters given in Table 3.1. **a** Description of the boundary value problem, **b** contour plots of the strain component ε_{11} and **c** profiles of the strain and stress components, ε_{11} and $\sigma_{22} = T_{22}$, across the film's thickness. After Cordero et al. (2016)

Table 3.1 Set of material parameters used in the finite element simulations, and the corresponding characteristic lengths and dimensionless parameters

E [GPa]	ν	A [N]	B [N.nm^2]	c_1 [N]	$c_{i\neq 1}$ [N]	b_0 [N]
140	0.3	$9.4\,10^{-8}$	$4.7\,10^{-8}$	$-2.8\,10^{-7}$	0	0

E [GPa]	ν	L_1 [nm]	L_2^2/L_1^2	η	η_0
140	0.3	1.0	0.5	−3	0

imposed between the top and bottom surfaces. The lateral surfaces $x_1 = \pm d/2$ are traction-free at any order, meaning that

$$t^1 = t^2 = t^3 = 0 \quad \text{at} \quad x_1 = \pm\frac{h}{2}, \tag{3.76}$$

at the vertical sides of the 2D geometry presented in Fig. 3.4. In these conditions, the nano-films can be considered of infinite length and finite width, d.

Tensile tests on thin films with various thickness values, d, are simulated in that way. In these tests, the nano-objects are assumed to remain elastic. The contour plots of Fig. 3.4 and the corresponding profiles across the thickness in Fig. 3.4 describe the simulated fields of the lateral strain and axial stress components, ε_{11} and T_{22}, obtained for three thicknesses and with the material parameters given in Table 3.1. These parameters are presented using the notation (3.51) introduced in section "Surface Energy Effects in Third Gradient Elasticity," and the corresponding generalized moduli A, B, c_i and b_0 are also given. The results given in Fig. 3.4 reveal that the surface stress effects are localized at the surface of the larger film of thickness $d = 50$ nm. The fields and profiles of ε_{11} and T_{22} are then close to those that can be obtained with a classical continuum. However, for smaller d, the surface stress effects become more significant and the affected zone spreads across the thickness. These results are similar to the previous observations made in the infinite strip shear case.

Recalling that plane strain conditions are used, the apparent Poisson ratio, ν_{app}, and apparent Young modulus, E^{app} are calculated as

$$\nu^{app} = \frac{\varepsilon_{11}}{\varepsilon_{11} - \varepsilon_{22}} \tag{3.77}$$

and

$$E^{app} = \frac{T_{22}(1 - \nu^{app^2})}{\varepsilon_{22}}, \tag{3.78}$$

where the macroscopic strain and stress ε_{11}, ε_{22} and T_{22} are computed from the finite element simulations by averaging the corresponding quantities over the whole sam-

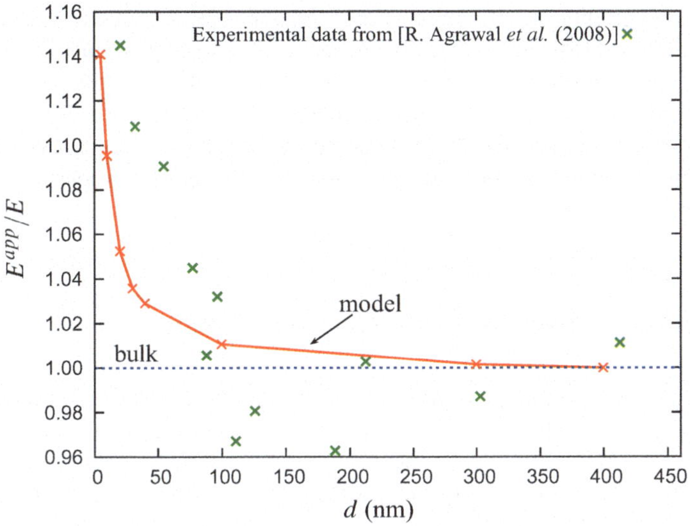

Fig. 3.5 Effect of the nano-film thickness, d, on the evolution of the apparent Young's modulus, E^{app}, for the material parameters given in Table 3.1. The horizontal line corresponds to the classical solution obtained without surface effects or equivalently to the bulk Young's modulus. The plotted experimental data come from Agrawal et al. (2008) and are obtained with tensile tests of single crystal ZnO nano-wires having a [0001] oriented wurtzite structure

ple. These apparent elastic constants are found to be size-dependent. Figure 3.5 shows the effect of the film thickness, d, on the evolution of the apparent Young's modulus, E^{app}, for the material parameters given in Table 3.1. These parameters were identified in order to reproduce, as quantitatively as possible, the size-dependent behavior of ZnO nano-wires for which experimental data are available (Agrawal et al. 2008). In the experiment, tensile tests were performed on single crystal ZnO nano-wires having a [0001] oriented wurtzite structure. It must be noted that the real single crystal exhibits anisotropic elastic and surface properties, whereas the present work concentrates on isotropic second strain gradient elasticity. Accordingly, the performed identification is valid only in the [0001] direction. Besides, fitting the simulation results for nano-plates to the experimental results for nano-wires can only lead to tentative and qualitative values of the corresponding moduli. The experimental results are far too scattered and not numerous enough to identify a unique set of material parameters. The chosen parameters provide sensible results, though probably exaggerated since the actual size effects may arise at smaller length scales than 100 nm predicted by the model in Fig. 3.5. A more accurate identification is possible by means of molecular dynamics or ab initio simulations (Shodja and Enzevaee 2019). However, large uncertainties remain also with this method because the confidence in higher-order moduli predicted by such discrete methods is necessarily limited.

The second gradient of strain theory presented in section "Second Strain Gradient Elasticity of Nano-Objects" is thus able to account, at least qualitatively, for the size-dependent elastic properties of nano-objects. Their mechanical behavior is affected by surface effects that occur in a surface region of a few nanometers. The obtained results show that these surface effects can be neglected when the thickness is equal to some tens of nanometers or larger. For such sizes, the elastic properties of the bulk material are retrieved.

Dispersion of Waves in Strain Gradient Media

The previous applications of third gradient elasticity were dedicated to the statics of fluids and solids. However, the strain gradient theories were conceived in the 1960s also to address the dispersion of elastic waves in heterogeneous materials including crystals and lattices (Toupin 1962). Homogeneous classical elasticity is known to be non-dispersive. In contrast, homogeneous gradient elasticity leads to dispersion effects when the wavelength becomes comparable to the intrinsic lengths of the medium defined as specific ratios of higher -order and classical moduli. This is a unique property which has been explored to identify the intrinsic lengths of gradient continua. Gradient media also possess intrinsic microinertia terms that are essential in the derivation of dispersion curves. Dispersion equations in gradient models are derived for instance in Askes et al. (2008b), Askes et al. (2008a), Maugin and Metrikine (2010). A discussion of the type of dispersion behavior depending on the type of generalized continuum theories can be found in Forest and Sab (2017). A weakness of the gradient theory is that the dispersion diagram contains only the acoustic branch. The optical branch cannot be predicted. In contrast, micromorphic theories of Eringen Eringen and Suhubi (1964) and Mindlin Mindlin (1964) predict both branches. This feature is similar to the analysis of vibrations in an Euler–Bernoulli beam, a gradient 1D model, and in a Timoshenko beam, a micromorphic-like (Cosserat) 1D model.

This remarkable dynamical response of gradient media has gained attention again very recently in the context of mechanical metamaterials (Nassar et al. 2016). The propagation of elastic waves in architectured materials, like lattices, has been studied numerically in Madeo et al. (2014), Rosi and Auffray (2016), Madeo et al. (2016), Rosi et al. (2018). These contributions reveal the anisotropy of wave propagation in hexagonal lattices, and the existence of band gaps in the dispersion diagram provided that a micromorphic model is applied.

Stress Gradient Elasticity

Much attention has been dedicated to strain gradient effects in continuum mechanics and materials sciences in the last fifty years, since the pioneering work of Toupin

(1962) and Mindlin (1965). In contrast, the role of stress gradients has been the subject of little attention if one excepts its introduction in fatigue crack initiation models at notches and holes of various sizes as studied in the engineering community (Bascoul and Maso 1981; Lahellec et al. 2005). More recently, a stress gradient-based criterion has been proposed for dislocation nucleation in crystals at a nanoscale (Miller and Acharya 2004).

Regarding continuum mechanics, there is a long-standing misconception or, at least, ambiguity going through the whole literature on generalized continua that implicitly considers that the strain gradient theory can also be regarded as a stress gradient model. The stress gradient can be found in Aifantis gradient elasticity model (Aifantis 1992; Ru and Aifantis 1993; Lazar et al. 2006; Aifantis 2009) in the form:

$$\boldsymbol{\tau} = \boldsymbol{T} - c\nabla^2 \boldsymbol{T} \tag{3.79}$$

where $\boldsymbol{\tau}$ is an effective stress tensor whose divergence vanishes in the absence of body forces and c is a material parameter associated with a characteristic length. In a Cartesian orthonormal coordinate system, the Laplace operator is applied to each component of the matrix. The Laplace term arises as the divergence of the gradient of the stress field. However, it can be shown that the presence of the stress gradient in this model is the result of a specific constitutive assumption made in Mindlin's strain gradient elasticity (Forest and Aifantis 2010). Accordingly, Aifantis gradient elasticity must be considered as a strain gradient model.

As a result, the question arises whether it is possible to formulate a stress gradient continuum theory describing size-dependent material properties and how much it may differ from the well-established strain gradient model.

Such a theory has been worked out by Forest and Sab (2012) and later on by Polizzotto (2014). The theory presented in Forest and Sab (2012) is reported in the present section.

Within the small strain framework (see Forest and Sab 2017 for the extension at large strains), the Cauchy simple stress tensor is

$$\boldsymbol{T} = T_{ij}\, \boldsymbol{e}_i \otimes \boldsymbol{e}_j \tag{3.80}$$

The stress gradient tensor is defined as

$$\boldsymbol{T} \otimes \nabla = T_{ij,k}\, \boldsymbol{e}_i \otimes \boldsymbol{e}_j \otimes \boldsymbol{e}_k \tag{3.81}$$

Its divergence is the vector $\boldsymbol{T} \cdot \nabla = T_{ij,j}\, \boldsymbol{e}_i$.

The space of third rank tensors that are symmetric with respect to the first two indices is denoted by $\mathcal{R}$. It is a vector space of dimension 18 which is endowed with the scalar product:

$$\boldsymbol{R} \;\therefore\; \boldsymbol{R} = R_{ijk} R_{ijk}, \quad \forall \boldsymbol{R} \in \mathcal{R} \tag{3.82}$$

Each tensor $\boldsymbol{R} \in \mathcal{R}$ can then be decomposed into a spherical part $\boldsymbol{R}^s \in \mathcal{S} \subset \mathcal{R}$ and a deviatoric part $\boldsymbol{R}^d \in \mathcal{D} \subset \mathcal{R}$:

$$\boldsymbol{R} = \boldsymbol{R}^s + \boldsymbol{R}^d \tag{3.83}$$

with

$$R^s_{ijk} = \frac{1}{4}\left(R_{ilm}\delta_{lm}\delta_{jk} + R_{jlm}\delta_{lm}\delta_{ik}\right) \tag{3.84}$$

Here, the space $\mathcal{D}$ is the subset of $\mathcal{R}$ containing the deviatoric elements $\boldsymbol{R}$ such that

$$\boldsymbol{R} : \mathbf{1} = 0, \quad R_{ijk}\delta_{jk} = 0 \tag{3.85}$$

The spherical part of the stress gradient is directly related to the divergence of the stress tensor by

$$(\boldsymbol{T} \otimes \nabla)^s_{ijk} = \frac{1}{4}(T_{im,m}\delta_{jk} + T_{jm,m}\delta_{ik}) \tag{3.86}$$

or equivalently,

$$(\boldsymbol{T} \otimes \nabla) : \mathbf{1} = (\boldsymbol{T} \otimes \nabla)^s : \mathbf{1} = \boldsymbol{T} \cdot \nabla \tag{3.87}$$

Construction of an Elastic Stress Gradient Material Theory

A homogeneous elastic Cauchy material is considered occupying the domain Ω. Clamping conditions are imposed at its boundary $\partial\Omega$ where the displacement vector vanishes: $\boldsymbol{u}(\boldsymbol{x}) = 0, \forall \boldsymbol{x} \in \partial\Omega$. The solid Ω is subjected to body forces $\boldsymbol{f}$. The variational formulation of the considered boundary value problem consists in minimizing the complementary energy

$$W^*(\boldsymbol{T}) = \int_\Omega \frac{1}{2}\boldsymbol{T} : \boldsymbol{S} : \boldsymbol{T}\, dv$$

where $\boldsymbol{S}$ is the elastic compliance tensor of fourth rank. The minimization takes place with respect to all statically admissible stress fields $\boldsymbol{T}$, i.e., fulfilling the balance equations.

The proposed stress gradient theory is based on the idea that the deviatoric part of the stress gradient can contribute to the energy, in addition to the stress tensor itself. The space SA^{SG} of statically admissible fields now contains the second rank tensor $\boldsymbol{T}$ and the third rank $\boldsymbol{R}$ that fulfill the following conditions on Ω:

$$\boldsymbol{T} \cdot \nabla + \boldsymbol{f} = 0 \tag{3.88}$$

$$\boldsymbol{R} = (\boldsymbol{T} \otimes \nabla)^d \tag{3.89}$$

Note that the spherical part of the stress gradient is entirely determined by the first balance equation (3.88) so that its deviatoric part only, denoted by $\boldsymbol{R}$, can enter the stress energy potential.

The stress energy density potential is $w^*(\boldsymbol{T}, \boldsymbol{R})$. The solution of the boundary value problem considered previously is now obtained by minimizing the complementary energy functional

$$W^{*SG}(\boldsymbol{T}) = \int_\Omega w^*(\boldsymbol{T}, \boldsymbol{R})\, dv \tag{3.90}$$

with respect to all $(\boldsymbol{T}, \boldsymbol{R}) \in SA^{SG}$.

To obtain the dual variational formulation of the previous stress gradient problem, we multiply the Eq. (3.88) by the displacement vector and integrate by parts over Ω:

$$\int_\Omega \boldsymbol{T} : \boldsymbol{\varepsilon}\, dv = \int_\Omega \boldsymbol{f} \cdot \boldsymbol{u}\, dv + \int_{\partial\Omega} (\boldsymbol{T} \cdot \boldsymbol{n}) \cdot \boldsymbol{u}\, da \tag{3.91}$$

where $\boldsymbol{\varepsilon}$ is the strain tensor defined as the symmetric part of the displacement gradient field.

In a similar way, we multiply the Eq. (3.89) by new kinematic variables $\boldsymbol{\Phi} \in \mathcal{D}$ of third rank and integrate over Ω. Integration by part leads to the variational formulation:

$$\int_\Omega \boldsymbol{T} : \boldsymbol{\Phi} \cdot \nabla + \boldsymbol{R} \,\vdots\, \boldsymbol{\Phi}\, dv = \int_{\partial\Omega} \boldsymbol{T} : \boldsymbol{\Phi} \cdot \boldsymbol{n}\, da \tag{3.92}$$

Summing up the Eqs. (3.91) and (3.92), the following variational formulation is obtained

$$\int_\Omega \left(\boldsymbol{T} : (\boldsymbol{\varepsilon} + \boldsymbol{\Phi} \cdot \nabla) + \boldsymbol{R} \,\vdots\, \boldsymbol{\Phi} \right) dv = \int_\Omega \boldsymbol{f} \cdot \boldsymbol{u}\, dv + \int_{\partial\Omega} \boldsymbol{T} : (\boldsymbol{u} \overset{s}{\otimes} \boldsymbol{n} + \boldsymbol{\Phi} \cdot \boldsymbol{n})\, da \tag{3.93}$$

This expression provides the definition of the strain measures that are conjugate to the stress and stress gradient tensors. The generalized strain measure is

$$\boldsymbol{e} := \boldsymbol{\varepsilon} + \boldsymbol{\Phi} \cdot \nabla \tag{3.94}$$

whereas the new kinematic degrees of freedom $\boldsymbol{\Phi}$ are conjugate to $\boldsymbol{R}$. These new kinematic d.o.f. are called *micro-displacements* and have the physical dimension of length. It is recalled that both $\boldsymbol{\Phi}$ and $\boldsymbol{R}$ are deviatoric third-order tensors. The conjugate stress and strain measures are related by the constitutive equations:

$$\boldsymbol{e} = \frac{\partial w^*(\boldsymbol{T}, \boldsymbol{R})}{\partial \boldsymbol{T}} \quad \text{and} \quad \boldsymbol{\Phi} = \frac{\partial w^*(\boldsymbol{T}, \boldsymbol{R})}{\partial \boldsymbol{R}}$$

The variational formulation (3.93) also provides the new clamping conditions for the considered boundary value problem:

$$\boldsymbol{u} \overset{s}{\otimes} \boldsymbol{n} + \boldsymbol{\Phi} \cdot \boldsymbol{n} = 0, \quad u_{(i}n_{j)} + \Phi_{ijk}n_k = 0 \tag{3.95}$$

Existence and uniqueness theorems for the corresponding boundary value problem in stress gradient elasticity have been established by Sab et al. (2016).

Method of Virtual Power for the Stress Gradient Medium

The balance equations and general boundary conditions for stress gradient media can also be formulated independently of the type of material behavior. This is done now by means of the method of virtual power as used by Germain in the case of strain gradient media (Germain 1973a; Maugin 1980).

Derivation of Balance Equations and Boundary Conditions

The usual set of kinematic degrees of freedom (DOF) available at each material point, namely the components of the displacement vector, is extended to incorporate the components of an element $\boldsymbol{\Phi} \in \mathcal{D}$. The two sets of kinematic DOF are assumed to be independent and to both have the dimension of length. The new kinematic DOF Φ_{ijk} are symmetric with respect to the first two indices. They are deviatoric in the sense defined in section "Construction of an Elastic Stress Gradient Material Theory" for third rank tensors belonging to $\mathcal{D}$. They are called *micro-displacements*, by analogy to the micro-deformation degrees of freedom introduced in the micromorphic theory by Eringen (1999). In the 3D case, the number of degrees of freedom of the theory at a material point is 18, corresponding to three components of displacements and 15 micro-displacement components, due to the constraint that $\boldsymbol{\Phi}^s = 0$.

The proposed theory is a first gradient theory with respect to the sets of degrees of freedom. The gradient of the displacement is assumed to work with the usual symmetric Cauchy stress. Furthermore, we assume that only the divergence of the micro-displacement also contributes to the overall work. The following form of the virtual work density of internal forces is postulated, as a linear form with respect to the degrees of freedom and their first gradients:

$$p^{(i)}(\boldsymbol{u}^*, \boldsymbol{\Phi}^*) = \boldsymbol{T} : \boldsymbol{u}^* \otimes \nabla + \boldsymbol{R} \,\vdots\, \boldsymbol{\Phi}^* + \boldsymbol{T} : \boldsymbol{\Phi}^* \cdot \nabla \qquad (3.96)$$

where $\boldsymbol{u}^*$ and $\boldsymbol{\Phi}^*$ are virtual displacement and micro-displacement fields. The stress conjugate to the gradient of displacement field is the symmetric Cauchy stress. We have assumed that the generalized stress conjugate to the divergence of the micro-displacement tensor is equal to the Cauchy stress itself. The generalized stress $\boldsymbol{R} \in \mathcal{D}$ is conjugate to the micro-displacement tensor and shares the same symmetry properties. The virtual work of internal forces on any sub-domain $V \subset \Omega$ is defined below and integrated by parts:

$$\mathcal{P}^{(i)}(\boldsymbol{u}^*, \boldsymbol{\Phi}^*) = \int_V p^{(i)}(\boldsymbol{u}^*, \boldsymbol{\Phi}^*)\, dv = \int_{\partial V} \boldsymbol{T} : (\boldsymbol{u}^* \overset{s}{\otimes} \boldsymbol{n} + \boldsymbol{\Phi} \cdot \boldsymbol{n})\, da$$
$$- \int_V \left((\boldsymbol{T} \cdot \nabla) \cdot \boldsymbol{u}^* + ((\boldsymbol{T} \otimes \nabla)^d - \boldsymbol{R}) \,\vdots\, \boldsymbol{\Phi} \right) dv \tag{3.97}$$

This prompts us to introduce the following form for the work density of contact forces at material points on the boundary ∂V, with normal vector $\boldsymbol{n}$:

$$p^{(c)}(\boldsymbol{u}^*, \boldsymbol{\Phi}^*) = \overset{2}{\boldsymbol{t}} : (\boldsymbol{u}^* \overset{s}{\otimes} \boldsymbol{n} + \boldsymbol{\Phi}^* \cdot \boldsymbol{n}) = t_{ij}(u_i^* n_j + \Phi_{ijk}^* n_k) \tag{3.98}$$

where $\overset{2}{\boldsymbol{t}}$ is a second-order symmetric stress tensor prescribed at the surface. The work density of forces acting at a distance is then

$$p^{(e)}(\boldsymbol{u}^*, \boldsymbol{\Phi}^*) = \boldsymbol{f} \cdot \boldsymbol{u}^* + \boldsymbol{F} \,\vdots\, \boldsymbol{\Phi}^* \tag{3.99}$$

where $\boldsymbol{f}$ and $\boldsymbol{F}$ are given first rank and third rank volume simple forces, respectively.

In the static case, the principle of virtual work then stipulates that, for every sub-domain V and for all virtual fields $\boldsymbol{u}^*$ and $\boldsymbol{\Phi}^*$, we have

$$\int_V p^{(i)}(\boldsymbol{u}^*, \boldsymbol{\Phi}^*)\, dv = \int_V p^{(e)}(\boldsymbol{u}^*, \boldsymbol{\Phi}^*)\, dv + \int_{\partial V} p^{(c)}(\boldsymbol{u}^*, \boldsymbol{\Phi}^*)\, da \tag{3.100}$$

Integration by parts and application of the principle for all sub-domains lead to the field equations of the problem

$$\boldsymbol{T} \cdot \nabla + \boldsymbol{f} = 0, \quad (\boldsymbol{T} \otimes \nabla)^d - \boldsymbol{R} + \boldsymbol{F} = 0, \quad \forall \boldsymbol{x} \in V \tag{3.101}$$

and the corresponding boundary conditions:

$$\overset{2}{\boldsymbol{t}} = \boldsymbol{T}, \quad \forall \boldsymbol{x} \in \partial V \tag{3.102}$$

As a result, the stress gradient theory allows to prescribe all the components of the stress tensor at the boundary. There are six dual conditions which amount to fixing the six kinematic components of $\boldsymbol{u} \overset{s}{\otimes} \boldsymbol{n} + \boldsymbol{\Phi} \cdot \boldsymbol{n}$ at the boundary.

At this stage, one may ask why the theory has 18 independent degrees of freedom and only six boundary conditions. This is due to the fact that once the six components of the stress tensor T_{ij} are known at the boundary, the 12 independent components of the tangential derivative of the stress are automatically known.

Comparison with Mindlin's Strain Gradient Model

A parallel can be drawn between Mindlin's original strain gradient model and the proposed stress gradient theory. Fundamental differences arise. The pros and the cons regarding possible computational advantages of the stress gradient model are pointed out.

The stress gradient model introduces 15 more degrees of freedom than the strain gradient one. Both theories are based on one second rank and one third rank stress tensors but in the stress gradient model the third rank generalized stress tensor is deviatoric. Both theories require six boundary conditions to be prescribed. The six essential conditions in strain gradient theory amount to fixing the components of the displacement and normal gradient vector of the displacement. The corresponding essential conditions in the stress gradient case deal with the six components $u_{(i}n_{j)} + \Phi_{ijk}n_k$. The natural conditions of the stress gradient model consist in prescribing the six stress components. In contrast, the natural conditions of the strain gradient model are quite intricate since the traction vector depends on the surface curvature and on the tangent derivative of the double stress tensor. The question arises of the mechanical meaning of the micro-displacement degrees of freedom. It is largely open, but first micromechanical definitions from the underlying microstructure can be found in recent contributions (Sab and Lebée 2015; Tran et al. 2018; Hütter et al. 2019).

Relation to Germain's General Micromorphic Theory

It is instructive to compare the previous gradient theories to one of the most general theories of materials with microstructure, namely Germain's general micromorphic continuum model (Germain 1973b). This generalization of Eringen's micromorphic model (Eringen 1999) relies on the introduction of additional degrees of freedom in the form of second-order, third-order and possibly higher-order deformation tensors. Let us concentrate on the second-order micromorphic theory Germain proposed, which is based on the following form of the power density of internal forces:

$$\begin{aligned} p^{(i)} &= \boldsymbol{T} : \boldsymbol{u} \otimes \nabla + \overset{2}{\boldsymbol{s}}{}^{\chi} : \boldsymbol{\chi} + \overset{3}{\boldsymbol{s}}{}^{\Phi} \vdots \boldsymbol{\Phi} + \overset{3}{\boldsymbol{S}}{}^{\chi} \vdots \boldsymbol{\chi} \otimes \nabla + \overset{4}{\boldsymbol{S}}{}^{\Phi} :: \boldsymbol{\Phi} \otimes \nabla \\ &= T_{i,j} u_{i,j} + s^{\chi}_{ij}\, \chi_{ij} + s^{\Phi}_{ijk}\, \Phi_{ijk} + S^{\chi}_{ijk}\, \chi_{ij,k} + S^{\Phi}_{ijkl}\, \Phi_{ijk,l} \end{aligned} \quad (3.103)$$

where the additional degrees of freedom χ_{ij} and Φ_{ijk} are independent second- and third-order tensor fields. These tensors are generally non-compatible fields, meaning that they do not correspond to gradients of a vector and second-order tensor fields, respectively. As a result, they generally do not display any symmetry property with respect to their indices. They are associated with the conjugate stresses $\overset{2}{\boldsymbol{s}}{}^{\chi}$, $\overset{3}{\boldsymbol{s}}{}^{\Phi}$, $\overset{3}{\boldsymbol{S}}{}^{\chi}$, $\overset{4}{\boldsymbol{S}}{}^{\Phi}$ in the power linear form. The principle of virtual power can be used to derive the balance equations fulfilled by the generalized stresses, written here in the static case in the absence of volume forces for the sake of brevity:

$$\boldsymbol{T} \cdot \nabla = 0, \quad \overset{3}{\boldsymbol{S}}{}^{\chi} \cdot \nabla - \overset{2}{\boldsymbol{s}}{}^{\chi} = 0, \quad \overset{4}{\boldsymbol{S}}{}^{\Phi} \cdot \nabla - \overset{3}{\boldsymbol{s}}{}^{\Phi} = 0 \tag{3.104}$$

At this stage, two special cases can be considered:

1. *First-order micromorphic model with kinematic internal constraint.* In the micromorphic model obtained when $\overset{3}{\boldsymbol{\Phi}} \equiv 0$, the following kinematic internal constraint is enforced:

$$\boldsymbol{\chi} \equiv \boldsymbol{u} \otimes \nabla, \quad \chi_{ij} \equiv u_{i,j} \tag{3.105}$$

which means that the micro-deformation $\boldsymbol{\chi}$ is no longer independent of the displacement field. Accordingly, the micro-deformation gradient $\boldsymbol{\chi} \otimes \nabla$ coincides with the second gradient of the displacement field $\boldsymbol{u} \otimes \nabla \otimes \nabla$. Mindlin's second gradient theory is retrieved in that way. The simple force stress tensor then is $\boldsymbol{T} + \boldsymbol{s}^{\chi}$, and Mindlin's hyperstress tensor is $\overset{3}{\boldsymbol{S}}{}^{\chi}$.

2. *Second-order micromorphic model with static internal constraint.* In the spirit of Eringen and Germain, $\boldsymbol{\chi}$ and $\boldsymbol{\Phi}$ have the physical dimensions of strain and strain gradient, respectively. But more general physical types of degrees of freedom can be imagined. For instance, we can enforce the following static internal constraint for the highest-order stress tensor:

$$\overset{4}{\boldsymbol{S}}{}^{\Phi} \equiv \boldsymbol{T} \otimes \boldsymbol{1}, \quad S^{\Phi}_{ijkl} \equiv T_{ij}\delta_{kl} \tag{3.106}$$

which means that the second-order stress tensor is not independent from the simple stress tensor. From the third balance equations in (3.104), it follows that

$$\overset{3}{\boldsymbol{s}}{}^{\Phi} = \overset{4}{\boldsymbol{S}}{}^{\Phi} \cdot \nabla = \boldsymbol{T} \otimes \nabla, \quad s^{\Phi}_{ijk} = S^{\Phi}_{ijkl,l} = (T_{ij}\delta_{kl})_{,l} = T_{ij,k} \tag{3.107}$$

Under these conditions, and considering the second-order micromorphic degrees of freedom only, the power density function (3.103) reduces to the expression (3.96) postulated for the stress gradient medium.

There are therefore strong links between constrained micromorphic models and stress and strain gradient theories. Considerations of internal constraints in generalized continua are discussed in the following references (Bertram and Glüge 2016; Bersani et al. 2019).

Simple Tension in Stress Gradient Elasticity

The case of simple tension is particularly straightforward and illustrative of the main features of the new model. In the idealized two-dimensional case of a bar (width $2H$) in tension along the Cartesian direction 1, the single non-vanishing component of the stress tensor is $T_{11}(x_2)$ taken as a function of the transverse coordinate x_2 along direction 2. The stress tensor cannot be homogeneous since it must vanish at the free lateral surface according to the generalized Neumann boundary conditions (3.102). It follows that the single non-vanishing stress gradient component is $R_{112} = T_{11,2}$. In that case, the stress gradient tensor fulfills the condition $R_{ijj} = 0$ as required by the balance of momentum equation. The axial strain $\varepsilon_{11} = \bar{\varepsilon} = u_{1,1}$ is homogeneous and prescribed as $\bar{\varepsilon}$.

The following simplified linear elastic law is assumed in the following:

$$\boldsymbol{\Phi} = E^{-1}\ell^2\, \boldsymbol{R} \tag{3.108}$$

where E is the Young modulus and ℓ is an intrinsic length arising in the stress gradient elasticity model. In the investigated tensile case, this gives

$$\Phi_{112} = E^{-1}\ell^2 R_{112} = E^{-1}\ell^2 T_{11,2} \tag{3.109}$$

The generalized strain measure (3.94) writes

$$e_{11} = \bar{\varepsilon} + \Phi_{112,2} = T_{11}/E \tag{3.110}$$

The last equation results from the application of the isotropic elasticity relation between the stress tensor and the generalized strain measure. The combination of the latter equation and the generalized Hooke's law (3.108) leads to the following differential equation

$$T_{11} - \ell^2 T_{11,22} = E\bar{\varepsilon} \tag{3.111}$$

to be solved for the stress T_{11}. Its solution is of the form

$$T_{11} = E\bar{\varepsilon} + A\cosh\frac{x_2}{\ell} + B\sinh\frac{x_2}{\ell} \tag{3.112}$$

The integration constants A and B are determined from the following boundary conditions:

$$T_{11}(x_2 = \pm H) = 0 \tag{3.113}$$

Finally, we obtain

$$T_{11} = E\bar{\varepsilon}\left(1 - \frac{\cosh(x_2/\ell)}{\cosh(H/\ell)}\right) \tag{3.114}$$

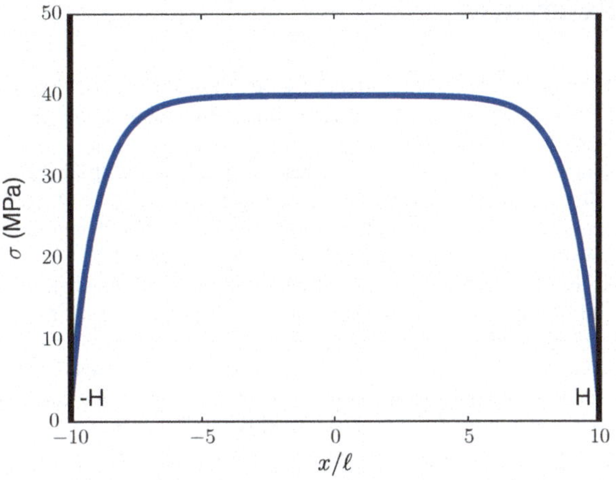

Fig. 3.6 Tensile test in stress gradient elasticity: stress profile in the cross section. The width of the bar is $2H$

The corresponding stress profile in a special case ($E = 200$ GPa, $\bar{\varepsilon} = 2.10^{-4}$) is represented by the curve of Fig. 3.6a.

Conclusions

Strain gradient elasticity can be used to account for size effects at the nanoscale in the presence of strong strain gradients that develop for instance in boundary layers and at free surfaces. Other applications exist in the mechanics of composite materials when the usual assumption of scale separation is not satisfied. This is the case when the size of fibers or other reinforcements becomes comparable to the wavelength of the applied loading conditions. Homogenization methods can then be extended to incorporate higher-order strain gradients and identify higher-order elasticity moduli; see Forest and Trinh (2011), Yvonnet (2019).

Structural analysis using the strain gradient continuum is not an easy task. Several implementations have been proposed in finite element codes using additional degrees of freedom and Lagrange multipliers in Shu et al. (1999) or based on the micromorphic continuum (Cordero et al. 2016).

Thermoelasticity of strain gradient media has been investigated to a lesser extent (Bertram and Forest 2014). The question of the presence of the temperature gradient or entropy gradient as a variable of the free/internal energy functions has been raised in Forest et al. (2000), Forest and Amestoy (2008), Khakalo and Niiranen (2020) but essentially remains open for applications involving high rates of strain and temperature gradients.

References

Agrawal, R., Peng, B., Gdoutos, E. E., & Espinosa, H. D. (2008). Elasticity Size Effects in ZnO Nanowires- A Combined Experimental-Computational Approach. *Nano Letters*, *8*(11), 3668–3674.

Aifantis, E. C. (1992). On the role of gradients on the localization of deformation and fracture. *International Journal of Engineering Science*, *30*, 1279–1299.

Aifantis, E. C. (2009). Deformation and failure of bulk nanograined and ultrafine-grained materials. *Materials Science and Engineering A*, *503*, 190–197.

Askes, H., Metrikine, A. V., Pichugin, A. V., & Bennett, T. (2008a). Four simplified gradient elasticity models for the simulation of dispersive wave propagation. *Philosophical Magazine*, *88*, 3415–3443.

Askes, H., Wang, B., & Bennett, T. (2008b). Element size and time step selection procedures for the numerical analysis of elasticity with higher-order inertia. *Journal of Sound and Vibration*, *314*, 650–656. https://doi.org/10.1016/j.jsv.2007.12.034.

Bascoul, A., & Maso, J. C. (1981). Influence des gradients de contraintes ou déformations sur le seuil d'écoulement plastique d'un acier. *Journal de Mécanique Appliquée*, *5*, 375–403.

A. Bersani, F. dell'Isola, and P. Seppecher. Lagrange multipliers in infinite dimensional spaces, examples of application. working paper or preprint, 2019. URL https://hal.archives-ouvertes.fr/hal-02269765.

Bertram, A., & Forest, S. (2014). The thermodynamics of gradient elastoplasticity. *Continuum Mechanics and Thermodynamics*, *26*, 269–286.

Bertram, A., & Glüge, R. (2016). Gradient materials with internal constraints. *Mathematics and Mechanics of Complex Systems*, *4*, 1–15.

Casal, P. (1961). La capillarité interne. *Cahier du Groupe Français d'Etudes de Rhéologie*, *6*, 31–37.

Casal, P. (1963). Capillarité interne en mecanique. *CR Acad. Sci. Paris*, *256*, 3820–3822.

Casal, P. (1972). La théorie du second gradient et la capillarité. *CR Acad. Sci. Paris*, *274*, 1571–1574.

Casal, P., & Gouin, H. (1985). Relation entre l'équation de l'énergie et l'équation du mouvement en théorie de Korteweg de la capillarité. *Comptes-rendus des séances de l'Académie des sciences. Série 2, Mécanique-physique, chimie, sciences de l'univers, sciences de la terre*, 300:231–234.

Cordero, N. M., Forest, S., & Busso, E. P. (2016). Second strain gradient elasticity of nano-objects. *Journal of the Mechanics and Physics of Solids*, *97*, 92–124. https://doi.org/10.1016/j.jmps.2015.07.012.

Dingreville, R., Qu, J., & Cherkaoui, M. (2005). Surface free energy and its effect on the elastic behavior of nano-sized particles, wires and films. *Journal of the Mechanics and Physics of Solids*, *53*, 1827–1854.

Dreyer, W., & Duderstadt, F. (2006). On the Becker/Döring theory of nucleation of liquid droplets in solids. *Journal of Statistical Physics*, *123*, 55–87.

Eringen, A. C. (1999). *Microcontinuum field theories*. New York: Springer.

Eringen, A. C., & Suhubi, E. S. (1964). Nonlinear theory of simple microelastic solids. *Int. J. Engng Sci.*, *2*(189–203), 389–404.

Forest, S., & Aifantis, E. C. (2010). Some links between recent gradient thermo-elasto-plasticity theories and the thermomechanics of generalized continua. *International Journal of Solids and Structures*, *47*, 3367–3376.

Forest, S., & Amestoy, M. (2008). Hypertemperature in thermoelastic solids. *Comptes Rendus Mécanique*, *336*, 347–353.

Forest, S., & Sab, K. (2012). Continuum stress gradient theory. *Mechanics Research Communications*, *40*, 16–25.

Forest, S., & Sab, K. (2017). Finite deformation second order micromorphic theory and its relations to strain and stress gradient models. *Mathematics and Mechanics of Solids*. https://doi.org/10.1177/1081286517720844.

Forest, S., & Trinh, D. K. (2011). Generalized continua and non-homogeneous boundary conditions in homogenization methods. *ZAMM Zeitschrift fuer Angewandte Mathematik und Mechanik*, *91*, 90–109.

Forest, S., Cardona, J.-M., & Sievert, R. (2000). Thermoelasticity of second-grade media. In G. A. Maugin, R. Drouot, & F. Sidoroff (Eds.), *Continuum thermomechanics* (pp. 163–176)., The art and science of modelling material behaviour, Paul Germain's anniversary volume Dordrecht: Kluwer Academic Publishers.

Fried, E., & Gurtin, M. E. (2006). Tractions, balances, and boundary conditions for nonsimple materials with application to liquid flow at small-length scales. *Archive for Rational Mechanics and Analysis*, *182*, 513–554.

Germain, P. (1973a). La méthode des puissances virtuelles en mécanique des milieux continus, première partie : théorie du second gradient. *Journal de Mécanique*, *12*, 235–274.

Germain, P. (1973b). The method of virtual power in continuum mechanics. Part 2 : Microstructure. *SIAM Journal on Applied Mathematics*, *25*, 556–575.

Gurtin, M. E. (1965). Thermodynamics and the possibility of spatial interaction in elastic materials. *Archive for Rational Mechanics and Analysis*, *19*, 339–352.

Gurtin, M. E., & Murdoch, A. I. (1978). Surface Stress in Solids. *International Journal of Solids and Structures*, *14*, 431–440. https://doi.org/10.1016/0020-7683(78)90008-2.

Hütter, G., Sab, K., & Forest, S. (2019). Kinematics and constitutive relations in the stress-gradient theory: interpretation by homogenization. arXiv:1906.08126.

Javili, A., Ottosen, N. S., Ristinmaa, M., & Mosler, J. (2018). Aspects of interface elasticity theory. *Mathematics and Mechanics of Solids*, *23*, 1004–1024. https://doi.org/10.1177/1081286517699041.

Khakalo, S., & Niiranen, J. (2018). Form ii of mindlin's second strain gradient theory of elasticity with a simplification: For materials and structures from nano- to macro-scales. *European Journal of Mechanics - A/Solids*, *71*, 292–319. https://doi.org/10.1016/j.euromechsol.2018.02.013.

Khakalo, S., & Niiranen, J. (2020). Anisotropic strain gradient thermoelasticity for cellular structures: Plate models, homogenization and isogeometric analysis. *Journal of the Mechanics and Physics of Solids*, *134*, 103728. https://doi.org/10.1016/j.jmps.2019.103728.

Lahellec, N., Bordreuil, C., & Hochard, C. (2005). Etude de la rupture fragile d'un stratifié quasi-isotrope à plis tissés: mise en évidence d'une longueur interne. *Comptes Rendus Mécanique*, *333*, 293–298.

Lazar, M., Maugin, G. A., & Aifantis, E. C. (2006). Dislocations in second strain gradient elasticity. *International Journal of Solids and Structures*, *43*, 1787–1817.

Madeo, A., Neff, P., Ghiba, I. D., Placidi, L., & Rosi, G. (2014). Wave propagation in relaxed micromorphic continua: Modeling metamaterials with frequency band-gaps. *Continuum Mechanics and Thermodynamics*, *27*, 551–570. https://doi.org/10.1007/s00161-013-0329-2.

Madeo, A., Barbagallo, G., d'Agostino, M. V., Placidi, L., & Neff, P. (2016). First evidence of non-locality in real band-gap metamaterials: Determining parameters in the relaxed micromorphic model. *Proceedings of the Royal Society of London A*, *472*, 20160169. https://doi.org/10.1098/rspa.2016.0169.

Maugin, G. A. (1980). The method of virtual power in continuum mechanics: Application to coupled fields. *Acta Mechanica*, *35*, 1–70.

Maugin, G. A., & Metrikine, A. V. (2010). *Mechanics of generalized continua, One hundred years after the Cosserats*. Advances in mechanics and mathematics (Vol. 21). Berlin: Springer.

McDowell, M. T., Leach, A. M., & Gall, K. (2008). Bending and tensile deformation of metallic nanowires. *Modelling and Simulation in Materials Science and Engineering*, *16*, 045003.

Miller, R. E., & Acharya, A. (2004). A stress-gradient based criterion for dislocation nucleation in crystals. *Journal of the Mechanics and Physics of Solids*, *52*, 1507–1525.

Mindlin, R. D. (1964). Micro-structure in linear elasticity. *Archive for Rational Mechanics and Analysis*, *16*, 51–78.

Mindlin, R. D. (1965). Second gradient of strain and surface-tension in linear elasticity. *International Journal of Solids and Structures*, *1*, 417–438.

Mindlin, R. D., & Eshel, N. N. (1968). On first strain gradient theories in linear elasticity. *International Journal of Solids and Structures*, *4*, 109–124.
Müller, P., & Saúl, A. (2004). Elastic effects on surface physics. *Surface Science Reports*, *54*, 157–258.
Nassar, H., He, Q.-C., & Auffray, N. (2016). A generalized theory of elastodynamic homogenization for periodic media. *International Journal of Solids and Structures*, *84*, 139–146. https://doi.org/10.1016/j.ijsolstr.2016.01.022.
Polizzotto, C. (2014). Stress gradient versus strain gradient constitutive models within elasticity. *International Journal of Solids and Structures*, *51*, 1809–1818. https://doi.org/10.1016/j.ijsolstr.2014.01.021.
Rocard, Y. (1967). *Thermodynamique*. Paris: Masson.
Rosi, G., & Auffray, N. (2016). Anisotropic and dispersive wave propagation within strain-gradient framework. *Wave Motion*, *63*, 120–134. https://doi.org/10.1016/j.wavemoti.2016.01.009.
Rosi, G., Placidi, L., & Auffray, N. (2018). On the validity range of strain-gradient elasticity: A mixed static-dynamic identification procedure. *European Journal of Mechanics - A/Solids*, *69*, 179–191. https://doi.org/10.1016/j.euromechsol.2017.12.005.
Ru, C. Q., & Aifantis, E. C. (1993). A simple approach to solve boundary value problems in gradient elasticity. *Acta Mechanica*, *101*, 59–68.
Sab, K., & Lebée, A. (2015). *Homogenization of heterogeneous thin and thick plates*. New Jersey: Wiley.
Sab, K., Legoll, F., & Forest, S. (2016). Stress gradient elasticity theory: Existence and uniqueness of solution. *Journal of Elasticity*, *123*, 179–201. https://doi.org/10.1007/s10659-015-9554-1.
Sadeghian, H., Yang, C. K., Goosen, J. F. L., Van der Drift, E., Bossche, A., French, P. J., et al. (2009). Characterizing size-dependent effective elastic modulus of silicon nanocantilevers using electrostatic pull-in instability. *Applied Physics Letters*, *94*, 221903.
Seppecher. P. (1996). *Les fluides de Cahn–Hilliard*. Habilitation thesis, Université du Sud Toulon Var.
Seppecher, P. (2000). Second gradient theory: Application to Cahn-Hilliard fluids. In G. A. Maugin, R. Drouot, & F. Sidoroff (Eds.), *Continuum thermomechanics* (pp. 379–388)., The art and science of modelling material behaviour, Paul Germain's anniversary volume Dordrecht: Kluwer Academic Publishers.
Shodja, H. M., & Enzevaee, C. (2019). Surface characterization of face-centered cubic crystals. *Mechanics of Materials*, *129*, 15–22. https://doi.org/10.1016/j.mechmat.2018.10.009.
Shu, J. Y., King, W. E., & Fleck, N. A. (1999). Finite elements for materials with strain gradient effects. *International Journal for Numerical Methods in Engineering*, *44*, 373–391.
Thomas, O., Ponchet, A., & Forest, S. (Eds.). (2011). *Mechanics of nano–objects* (p. 380). Presses des Mines. ISBN: 978–2911256–67–7.
Toupin, R. A. (1962). Elastic materials with couple stresses. *Archive for Rational Mechanics and Analysis*, *11*, 385–414.
Tran, V. P., Brisard, S., Guilleminot, J., & Sab, K. (2018). Mori-Tanaka estimates of the effective elastic properties of stress-gradient composites. *International Journal of Solids and Structures*, *146*, 55–68. https://doi.org/10.1016/j.ijsolstr.2018.03.020.
Truesdell, C., & Noll, W. (1965). *The non-linear field theories of mechanics*. Handbuch der Physik, edited by S. Flügge, Reedition. Berlin: Springer.
Yvonnet, J. (2019). *Computational homogenization of heterogeneous materials with finite elements*. Solid mechanics and its applications (Vol. 258). Berlin: Springer. ISBN 978-3-030-18382-0. https://doi.org/10.1007/978-3-030-18383-7_9.

Chapter 4
Microscopic Interpretation of Strain-Gradient and Generalized Continuum Models

Pierre Seppecher

Abstract Everybody has an intuitive interpretation of continuum mechanics in terms of microscopic forces: Many punctual forces applied to a material lead in a continuous approximation, that is, through a homogenization process, to a continuous surface density of forces applied on its boundary. Things are less clear when dealing with generalized continua or, worse, with strain-gradient models. For instance, interpretations of Cosserat variable are often proposed in terms of the rotation of some internal small substructures which are more rigid than the rest of the material; then, boundary actions are understood as a surface density of torques; however, it is seldom explained how these actions propagate through the body. In this chapter, we provide examples of lattices for which homogenization can be performed rigorously; they lead to generalized or strain-gradient continua; hence, the peculiarities of these models are enlightened by a possible microscopic interpretation.

Keywords Architectured material · Cosserat model · Couple-stress model · Cauchy theorem · Gamma-convergence · High contrast composite material · Intrinsic length · Meta-material · Pantographic structure · Strain gradient model · Variational formulation

Introduction: The Energetic Point of View

It is not so easy to define precisely what a strain-gradient material is. Simply assuming that the stress tensor depends on higher gradient of the displacement field means that the notion of stress has previously been defined and this generally means "defined through Cauchy theorem." When doing that, one implicitly accepts the fundamental assumption of Cauchy theorem that the mechanical interactions between two parts of the material are limited to a density of forces along the surface which divides the two parts. This way of introducing strain-gradient models is not coherent and leads to paradoxical results appearing here or there in the literature. Indeed, there

P. Seppecher (✉)
IMATH, Université de Toulon, 83130 La Garde cedex, France
e-mail: seppecher@univ-tln.fr

A. Bertram and S. Forest (eds.), *Mechanics of Strain Gradient Materials*,
CISM International Centre for Mechanical Sciences 600,
https://doi.org/10.1007/978-3-030-43830-2_4

are different notions of stresses which are the products of the considered continuum theories. Passing from a classical model to a strain-gradient one is more than a simple change of constitutive law.

In the general case, there are two ways to properly define a strain-gradient material: either making use of the virtual work (or virtual velocity) principle (see d'Alembert 1743; Germain 1973b; Dell'Isola et al. 2015) or applying an appropriate generalization of Cauchy theorem which is a much more cumbersome way (see dell'Isola and Seppecher 1997; Dell'Isola et al. 2015).

In the variational case, strain-gradient materials are much easier to define. That is why, in this chapter, we will only focus on the static case. The equilibrium displacement field $\boldsymbol{u}$ of an elastic body $\mathcal{B}$ subject to some bulk force $\boldsymbol{b}$ and to some surface forces $\boldsymbol{f}$ along its boundary $\partial\mathcal{B}$ minimizes the total energy made of the sum of the internal elastic energy and the external force potential.

Let us recall how that works in classical elasticity: The elastic energy $\mathcal{W}$ is the integral over the body of a density w which is a function of the strain tensor field. In this chapter, for the sake of simplicity, we will remain in the framework of small displacements and consider only a convex elastic energy. Denoting $E_{\boldsymbol{u}} = (\nabla\boldsymbol{u} + (\nabla\boldsymbol{u})^t)/2$ the linearized strain tensor, the elastic energy is written

$$\mathcal{W}(\boldsymbol{u}) := \int_{\mathcal{B}} w(E_{\boldsymbol{u}})\,\mathrm{d}V. \tag{4.1}$$

The equilibrium displacement $\boldsymbol{u}$ minimizes the total energy

$$\int_{\mathcal{B}} \big(w(E_{\boldsymbol{u}}) - \boldsymbol{b}\cdot\boldsymbol{u}\big)\,\mathrm{d}V - \int_{\partial\mathcal{B}} \boldsymbol{f}\cdot\boldsymbol{u}\,\mathrm{d}A$$

and therefore satisfies the variational formulation

$$\forall\,\delta\boldsymbol{u},\quad \int_{\mathcal{B}} \big(Dw(E_{\boldsymbol{u}})\cdot\cdot E_{\delta\boldsymbol{u}} - \boldsymbol{b}\cdot\delta\boldsymbol{u}\big)\,\mathrm{d}V - \int_{\partial\mathcal{B}} \boldsymbol{f}\cdot\delta\boldsymbol{u}\,\mathrm{d}A = 0.$$

The differential $Dw(E_{\boldsymbol{u}})$ of the energy density with respect to the strain can be represented by a symmetric tensor. Let us denote it by σ as usual. As $\sigma\cdot\cdot E_{\delta\boldsymbol{u}} = \sigma\cdot\cdot\nabla\delta\boldsymbol{u}$, we can integrate by parts the first addend (i.e., use the divergence theorem) and we get

$$\forall\,\delta\boldsymbol{u},\quad \int_{\mathcal{B}} \big(-\operatorname{div}(\sigma) - \boldsymbol{b}\big)\cdot\delta\boldsymbol{u}\,\mathrm{d}V + \int_{\partial\mathcal{B}} (\sigma\cdot\boldsymbol{n} - \boldsymbol{f})\cdot\delta\boldsymbol{u}\,\mathrm{d}A = 0$$

where $\boldsymbol{n}$ stands for the external normal to the boundary of the domain. This being true for any virtual field (or "test function") $\delta\boldsymbol{u}$, we deduce that $\boldsymbol{u}$ satisfies

$$\begin{cases} \operatorname{div}(\sigma) + \boldsymbol{b} = 0 & \text{on } \mathcal{B}, \\ \sigma = Dw(E_{\boldsymbol{u}}) & \text{on } \mathcal{B}, \\ \sigma\cdot\boldsymbol{n} = \boldsymbol{f} & \text{on } \partial\mathcal{B}. \end{cases}$$

We recover the classical system of equilibrium equations. A Dirichlet boundary condition, that is, a constraint $\boldsymbol{u} = \boldsymbol{u}_0$ upon the displacement on a part of the boundary, could easily be taken into account: The condition $\sigma \cdot \boldsymbol{n} = \boldsymbol{f}$ would then disappear on this part as the test functions $\delta \boldsymbol{u}$ would vanish there.

Definitions

In this context, defining a strain-gradient material is straightforward. It is enough to allow the elastic energy density to depend on the gradient of the strain field and to write

$$\mathcal{W}(\boldsymbol{u}) := \int_{\mathcal{B}} w(E_{\boldsymbol{u}}, \nabla E_{\boldsymbol{u}})\, \mathrm{d}V. \tag{4.2}$$

It is clear that any function of $E_{\boldsymbol{u}}$ and $\nabla E_{\boldsymbol{u}}$ is a function of $\nabla \boldsymbol{u}$ and $\nabla\nabla \boldsymbol{u}$. The converse is also true: $\nabla\nabla \boldsymbol{u}$ can be written in terms of $\nabla E_{\boldsymbol{u}}$, and thus any objective function of $(\nabla \boldsymbol{u}, \nabla\nabla \boldsymbol{u})$ that is function which vanishes for rigid motions can be written $w(E_{\boldsymbol{u}}, \nabla E_{\boldsymbol{u}})$. That is why *strain-gradient materials* and *second-gradient materials* are synonymous.

Defining a "generalized continuum" is as easy. We assume that the kinematics of the continuum is described by the displacement field $\boldsymbol{u}$ plus some extra kinematic field γ which may be of any tensorial nature. Then, we allow the elastic energy density to depend on the displacement gradient, on γ and on its gradient $\nabla\gamma$, writing

$$\mathcal{W}(\boldsymbol{u}) := \inf_{\gamma} \int_{\mathcal{B}} w(\nabla \boldsymbol{u}, \gamma, \nabla\gamma)\, \mathrm{d}V. \tag{4.3}$$

Allowing the density w in (4.1), (4.2) or (4.3) to take values in $[0, +\infty]$ is a synthetic way for taking some constraint into account. For instance, setting $w(E_{\boldsymbol{u}}) = +\infty$ in (4.1) as soon as the trace $tr(E_{\boldsymbol{u}})$ of the strain does not vanish clearly imposes to any minimizer to satisfy $tr(E_{\boldsymbol{u}}) = 0$ that is $\mathrm{div}(\boldsymbol{u}) = 0$: Incompressible elastic materials are thus described. Another important example for our concern is when the extra kinematic variable γ has the same tensorial nature as $\nabla \boldsymbol{u}$ and when one sets $w(\nabla \boldsymbol{u}, \gamma, \nabla\gamma) = +\infty$ in (4.3) as soon as $\gamma - \nabla \boldsymbol{u} \neq 0$. Any minimizer will then satisfy $\gamma = \nabla \boldsymbol{u}$, and its energy $w(\nabla \boldsymbol{u}, \gamma, \nabla\gamma) = w(\nabla \boldsymbol{u}, \nabla \boldsymbol{u}, \nabla\nabla \boldsymbol{u})$ becomes a strain-gradient one. This way of considering strain-gradient models as particular cases of generalized continua is preferred by many authors. This gives indeed interesting insights on the behavior of strain-gradient models, but it also hides some specificity of boundary conditions because it is not obvious to derive proper boundary conditions by using the divergence theorem when the energy density is allowed to take infinite values.

Of course, the variational point of view does not cover the majority of situations encountered in mechanics. The reader should consider it as a first step in the understanding of strain-gradient models, of their behavior, of their mechanical interactions on the boundary of the domain (or at any dividing surface inside the body), of the fact that equilibrium equations lead to well-posed mathematical problems and of the way microscopic forces can explain strain-gradient effects.

Well-Posedness and Approximation

The direct method of calculus of variations provides a standard way for checking whether the system of equilibrium equations has a solution. The idea, which is similar to what could be done in a numerical process, is to start with an admissible displacement field $\tilde{\boldsymbol{u}}$ and then to improve it (i.e., modify it in order to decrease its total energy) until convergence to equilibrium. One has simply to check that no "surprise" arises during this process:

- First check the existence of an admissible $\tilde{\boldsymbol{u}}$ ($\mathcal{W}(\tilde{\boldsymbol{u}}) < +\infty$). This means that one has not imposed incompatible constraints to the material. Once the existence of $\tilde{\boldsymbol{u}}$ is ensured, one knows that the infimum $\mathcal{W}_m := \inf_{\boldsymbol{u}}\{\mathcal{W}(\boldsymbol{u}) - \int_{\mathcal{B}} \boldsymbol{b} \cdot \boldsymbol{u}\, \mathrm{d}V\}$ belongs to $[-\infty, +\infty[$ and so that a minimizing sequence $(\boldsymbol{u}_n)$ exists (i.e., a sequence such that $\mathcal{W}(\boldsymbol{u}_n) - \int_{\mathcal{B}} \boldsymbol{b} \cdot \boldsymbol{u}_n\, \mathrm{d}V \to \mathcal{W}_m$).
- Then, check that $(\boldsymbol{u}_n)$ (or a subsequence) converges (in some sense): $\boldsymbol{u}_n \to \bar{\boldsymbol{u}}$. This is ensured by the coercivity of $\mathcal{W}$. This is a technical mathematical step which needs the choice of a suitable topology. The weaker is the chosen topology, the easiest is the convergence of $(\boldsymbol{u}_n)$.
- Finally, check that $\lim \mathcal{W}(\boldsymbol{u}_n) = \mathcal{W}(\lim \boldsymbol{u}_n)$ that is $\mathcal{W}_m = \mathcal{W}(\bar{\boldsymbol{u}})$. This is ensured by the (lower semi) continuity of $\mathcal{W}$. Again, the chosen topology is essential: It must be strong enough for ensuring this property.

At the end of the process, $\bar{\boldsymbol{u}}$ is an equilibrium solution and $\boldsymbol{u}_n$ is a sequence of approximations for it. All these steps work fine when w is a positive quadratic form, an assumption that we will assume from now on.

Equilibrium Equations

We have already recalled how one can derive the variational formulation and the system of equilibrium equations from the minimization of (4.1) in classical elasticity. Let us now see how to do it for strain-gradient materials and generalized continua.

Minimizing $\mathcal{W}(\boldsymbol{u}) - \int_{\mathcal{B}} \boldsymbol{b} \cdot \boldsymbol{u}\, \mathrm{d}V - \int_{\partial\mathcal{B}} \boldsymbol{f} \cdot \boldsymbol{u}\, \mathrm{d}A$ when $\mathcal{W}$ is Gateaux differentiable leads to the variational formulation $\forall\, \delta\boldsymbol{u}$, $D\mathcal{W}(\boldsymbol{u})(\delta\boldsymbol{u}) - \int_{\mathcal{B}} \boldsymbol{b} \cdot \delta\boldsymbol{u}\, \mathrm{d}V - \int_{\partial\mathcal{B}} \boldsymbol{f} \cdot \delta\boldsymbol{u}\, \mathrm{d}A = 0$. This reads, when $\mathcal{W}$ is given by (4.2),

$$\int_{\mathcal{B}} \big(\partial_1 w(E_{\boldsymbol{u}}, \nabla E_{\boldsymbol{u}})\cdot\cdot E_{\delta \boldsymbol{u}} + \partial_2 w(E_{\boldsymbol{u}}, \nabla E_{\boldsymbol{u}})\cdot\cdot\cdot \nabla E_{\delta \boldsymbol{u}} - \boldsymbol{b}\cdot\delta\boldsymbol{u}\big)\,\mathrm{d}V \\ - \int_{\partial\mathcal{B}} \boldsymbol{f}\cdot\delta\boldsymbol{u}\,\mathrm{d}A = 0.$$

Let us set $\overset{(1)}{T} := \partial_1 w(E_{\boldsymbol{u}}, \nabla E_{\boldsymbol{u}})$, $\overset{(2)}{T} := \partial_2 w(E_{\boldsymbol{u}}, \nabla E_{\boldsymbol{u}})$ and $T := \overset{(1)}{T} - \operatorname{div}(\overset{(2)}{T})$. We remark that $\overset{(1)}{T}$ is a symmetric tensor while $\overset{(2)}{T}$ is a third-order tensor symmetric with respect to the two first indices. Hence, T is also a symmetric matrix. Using this notation, the variational formulation reads

$$\int_{\mathcal{B}} \big(\overset{(1)}{T}\cdot\cdot E_{\delta \boldsymbol{u}} + \overset{(2)}{T}\cdot\cdot\cdot \nabla E_{\delta \boldsymbol{u}} - \boldsymbol{b}\cdot\delta\boldsymbol{u}\big)\,\mathrm{d}V - \int_{\partial\mathcal{B}} \boldsymbol{f}\cdot\delta\boldsymbol{u}\,\mathrm{d}A = 0.$$

To get the equilibrium equations, it is enough to consider smooth test functions $\delta\boldsymbol{u}$ and to integrate (twice) by parts (rigorously, this must be done in the sense of distributions). We successively obtain

$$\int_{\mathcal{B}} \Big((\overset{(1)}{T} - \operatorname{div}(\overset{(2)}{T}))\cdot\cdot E_{\delta \boldsymbol{u}} - \boldsymbol{b}\cdot\delta\boldsymbol{u}\Big)\,\mathrm{d}V \\ + \int_{\partial\mathcal{B}} \Big((\overset{(2)}{T}\cdot\boldsymbol{n})\cdot\cdot\nabla\delta\boldsymbol{u} - \boldsymbol{f}\cdot\delta\boldsymbol{u}\Big)\,\mathrm{d}A = 0,$$

$$\int_{\mathcal{B}} -\Big((\operatorname{div}(T) + \boldsymbol{b})\cdot\delta\boldsymbol{u}\Big)\,\mathrm{d}V \\ + \int_{\partial\mathcal{B}} \Big((\overset{(2)}{T}\cdot\boldsymbol{n})\cdot\cdot\nabla\delta\boldsymbol{u} + (T\cdot\boldsymbol{n} - \boldsymbol{f})\cdot\delta\boldsymbol{u}\Big)\,\mathrm{d}A = 0. \quad (4.4)$$

Considering functions $\delta\boldsymbol{u}$ with compact support in $\mathcal{B}$ gives the bulk equilibrium equation:

$$\operatorname{div}(T) + \boldsymbol{b} = 0. \quad (4.5)$$

Thus, the reader could infer that T plays the role of Cauchy stress tensor and that the equation

$$T := \overset{(1)}{T} - \operatorname{div}(\overset{(2)}{T}) = \partial_1 w(E_{\boldsymbol{u}}, \nabla E_{\boldsymbol{u}}) - \operatorname{div}(\partial_2 w(E_{\boldsymbol{u}}, \nabla E_{\boldsymbol{u}}))$$

is a constitutive law for it. As far as the bulk equation (which appears clearly to be here a fourth-order partial differential equation) is concerned, it is true. We will see that it is false when boundary conditions are concerned.

We remark that minimizing $\mathcal{W}$ provides (i) the order of the theory (classical or strain-gradient), (ii) constitutive laws for the hyper-stress tensors, (iii) the principle of virtual work and (iv) the bulk equilibrium equation.

The procedure is similar in the case of a generalized continuum. The minimization is performed with respect to both variables $\boldsymbol{u}$ and γ which are now independent. Thus, the variational formulation splits:

$$\begin{cases} \int_{\mathcal{B}} (\partial_1 w(\nabla \boldsymbol{u}, \gamma, \nabla\gamma)\cdot\cdot\nabla\delta\boldsymbol{u} - \boldsymbol{b}\cdot\delta\boldsymbol{u})\,\mathrm{d}V - \int_{\partial\mathcal{B}} \boldsymbol{f}\cdot\delta\boldsymbol{u}\,\mathrm{d}A = 0, \\ \int_{\mathcal{B}} (\partial_2 w(\nabla \boldsymbol{u}, \gamma, \nabla\gamma)\cdot\delta\gamma + \partial_3 w(\nabla \boldsymbol{u}, \gamma, \nabla\gamma)\cdot\cdot\nabla\delta\gamma)\,\mathrm{d}V = 0. \end{cases} \tag{4.6}$$

Let us set

$$\overset{(1)}{T} := \partial_1 w(\nabla \boldsymbol{u}, \gamma, \nabla\gamma),\quad \overset{(0)}{\tau} := \partial_2 w(\nabla \boldsymbol{u}, \gamma, \nabla\gamma) \text{ and } \overset{(1)}{\tau} := \partial_3 w(\nabla \boldsymbol{u}, \gamma, \nabla\gamma).$$

Remark that, unlike the case of strain-gradient materials, $\overset{(1)}{T}$ can now be a non-symmetric matrix. Indeed, we have not, at this point, fixed the tensorial nature of γ neither the way it depends on rigid motions. Using this notation, the variational formulation reads

$$\begin{cases} \int_{\mathcal{B}} \left(\overset{(1)}{T}\cdot\cdot\nabla\delta\boldsymbol{u} - \boldsymbol{b}\cdot\delta\boldsymbol{u}\right)\mathrm{d}V - \int_{\partial\mathcal{B}} \boldsymbol{f}\cdot\delta\boldsymbol{u}\,\mathrm{d}A = 0, \\ \int_{\mathcal{B}} \left(\overset{(0)}{\tau}\cdot\delta\gamma + \overset{(1)}{\tau}\cdot\cdot\nabla\delta\gamma\right)\mathrm{d}V = 0. \end{cases} \tag{4.7}$$

Integrating by parts both equations gives

$$\begin{cases} \int_{\mathcal{B}} \left((-\operatorname{div}(\overset{(1)}{T}) - \boldsymbol{b})\cdot\delta\boldsymbol{u}\right)\mathrm{d}V + \int_{\partial\mathcal{B}} (\overset{(1)}{T}\cdot\boldsymbol{n} - \boldsymbol{f})\cdot\delta\boldsymbol{u}\,\mathrm{d}A = 0, \\ \int_{\mathcal{B}} \left((-\operatorname{div}(\overset{(1)}{\tau}) + \overset{(0)}{\tau})\cdot\delta\gamma\right)\mathrm{d}V + \int_{\partial\mathcal{B}} (\overset{(1)}{\tau}\cdot\boldsymbol{n})\cdot\delta\gamma\,\mathrm{d}A = 0. \end{cases} \tag{4.8}$$

Considering smooth $\delta\boldsymbol{u}$ and $\delta\gamma$ with compact support in $\mathcal{B}$, we get the following system of bulk equilibrium equations (a coupled system of second-order partial differential equations)

$$\begin{cases} div(\overset{(1)}{T}) + b = 0, \\ div(\overset{(1)}{\tau}) - \overset{(0)}{\tau} = 0. \end{cases} \tag{4.9}$$

Boundary Conditions

Let us start by considering generalized continua. Incorporating bulk equilibrium Eq. (4.9) in the variational formulation (4.8) gives

$$\begin{cases} \int_{\partial\mathcal{B}} (\overset{(1)}{T} \cdot \boldsymbol{n} - \boldsymbol{f}) \cdot \delta\boldsymbol{u} \, \mathrm{d}A = 0, \\ \int_{\partial\mathcal{B}} (\overset{(1)}{\tau} \cdot \boldsymbol{n}) \cdot \delta\gamma \, \mathrm{d}A = 0 \end{cases} \tag{4.10}$$

which lead to the two boundary conditions

$$\overset{(1)}{T} \cdot \boldsymbol{n} = \boldsymbol{f} \quad \text{and} \quad \overset{(1)}{\tau} \cdot \boldsymbol{n} = 0 \quad \text{on } \partial\mathcal{B}. \tag{4.11}$$

Note that adding a term like $-\int_{\partial B} \Phi \cdot \gamma \, \mathrm{d}A$ in the total energy would make the second boundary condition non-homogeneous. The mechanical interpretation of the external action Φ would depend on the mechanical interpretation of the extra kinematic descriptor γ. For instance if γ is interpreted as a micro-rotation as in Cosserat theory, then Φ is an associated surface density of torques. Note that these boundary conditions can be replaced by dual conditions: If $\boldsymbol{u}$ or γ are fixed on some part of the boundary, then all variations $\delta\boldsymbol{u}$ or $\delta\gamma$ must vanish and we lost the associated boundary conditions there.

As far as strain-gradient materials are concerned, things are a bit more complex. Incorporating bulk equilibrium Eq. (4.5) in the variational formulation (4.4) gives

$$\int_{\partial\mathcal{B}} ((\overset{(2)}{T} \cdot \boldsymbol{n}) \cdot\cdot \nabla\delta\boldsymbol{u} + (T \cdot \boldsymbol{n} - \boldsymbol{f}) \cdot \delta\boldsymbol{u}) \, \mathrm{d}A = 0$$

but the values taken by $\delta\boldsymbol{u}$ and $\nabla\delta\boldsymbol{u}$ on the boundary of domain are not independent. The gradient of $\boldsymbol{u}$ on the boundary can be split in normal and tangent derivatives, and a new integration by parts on the manifold $\partial\mathcal{B}$ can be performed. Let us assume that the boundary of the domain is smooth.[1] We get[2]

$$\int_{\partial\mathcal{B}} \left((T \cdot \boldsymbol{n} - \mathrm{div}_{//}(\overset{(2)}{T} \cdot \boldsymbol{n})_{//} - \boldsymbol{f}) \cdot \delta\boldsymbol{u} + ((\overset{(2)}{T} \cdot \boldsymbol{n}) \cdot \boldsymbol{n}) \cdot (\nabla\delta\boldsymbol{u} \cdot n) \right) \mathrm{d}A = 0.$$

As $\delta\boldsymbol{u}$ and $\nabla\delta\boldsymbol{u} \cdot \boldsymbol{n}$ are independent quantities along the boundary, we obtain the two boundary conditions:

$$T \cdot \boldsymbol{n} - \mathrm{div}_{//}(\overset{(2)}{T} \cdot \boldsymbol{n})_{//} = \boldsymbol{f} \quad \text{and} \quad \overset{(2)}{T} \cdot \boldsymbol{n} = 0 \quad \text{on } \partial\mathcal{B}. \tag{4.12}$$

We see that T cannot be interpreted as the Cauchy stress tensor as it does not, by itself, represent the surface density of applied forces which is the basic assumption of Cauchy stress theory. It is also important to notice that the surface density of

[1]If the domain $\mathcal{B}$ is only piecewise smooth, integrating by parts on the smooth parts leads to concentration of forces along the edges of the domain (see Germain 1973b; dell'Isola and Seppecher 1997; dell'Isola et al. 2012).

[2]The tangent part of a vector field $\boldsymbol{V}$ on the boundary is denoted $\boldsymbol{V}_{//} := (Id - \boldsymbol{n} \otimes \boldsymbol{n}) \cdot \boldsymbol{V}$, while its surface divergence is denoted $\mathrm{div}_{//}(\boldsymbol{V}) := \nabla\boldsymbol{V} \cdot\cdot (Id - \boldsymbol{n} \otimes \boldsymbol{n})$.

applied forces does not depend linearly on the normal $\boldsymbol{n}$ but that a part of it depends linearly on the curvature of the surface (through the term $\nabla_{//}(\boldsymbol{n})_{//}$ which appears when developing the surface divergence). Hence, strain-gradient materials are not only incompatible with Cauchy postulate but also with the weaker assumption of (Noll, 1973) that the surface density of forces is uniformly bounded whatever the shape of the surface. Note that adding a term like $-\int_{\partial B} \Phi \cdot (\nabla \delta \boldsymbol{u} \cdot \boldsymbol{n})\, \mathrm{d}A$ in the total energy would make the second boundary condition non-homogeneous. The mechanical interpretation of the external action Φ (double force in the nomenclature of Germain 1973b) is not obvious. It is one of the goals of this chapter to give a microscopic interpretation of such an action. Remark again that these boundary conditions can be replaced by dual conditions: If $\boldsymbol{u}$ or $\nabla \boldsymbol{u} \cdot \boldsymbol{n}$ are fixed on some part of the boundary, then all variations must vanish and we lost the associated boundary conditions. There exist different types of "Dirichlet boundary conditions" on $\partial\mathcal{B}$ for these materials: a strong one which fixes $\boldsymbol{u} = \boldsymbol{u}_0$ and $\nabla \boldsymbol{u} \cdot \boldsymbol{n} = \boldsymbol{r}_0$, a weak one which only fixes $\boldsymbol{u} = \boldsymbol{u}_0$ and let free the normal derivative $\nabla \boldsymbol{u} \cdot \boldsymbol{n}$ and a last one fixing $\nabla \boldsymbol{u} \cdot \boldsymbol{n} = \boldsymbol{r}_0$ and letting $\boldsymbol{u}$ free.

As we already mentioned, it is not obvious to derive the boundary conditions (4.12) for strain-gradient models from the boundary conditions (4.11) of generalized continua by simply imposing the constraint $\nabla \boldsymbol{u} = \gamma$. This difficulty has led to some misunderstanding of the nature of mechanical interactions in strain-gradient models.

The Need for Microscopic Models

Since the very beginning of continuum mechanics, microscopic interpretations have been essential for getting a good grasp of continuous concepts. In particular, surface density of forces has been interpreted as the mean value of many discrete forces at the atomic range. Of course, it is inconceivable to try to justify all continuous models as limits of atomic ones. But it is useful to have at our disposal some toy models for getting intuition about the principal concepts of the theory.

For strain-gradient models, natural questions are: (i) How can a material react to the variations in space of the strain field, and (ii) what type of micro-forces can result in macroscopic mechanical interactions on the boundary of type $\overset{(2)}{T} \cdot \boldsymbol{n}$?

For generalized continua, questions are: (i) What is the microscopic interpretation of the extra kinematic variable γ, (ii) why does the material react to the variations in space of this variable, (iii) what is the reason of the coupling between this variable and the gradient of the displacement field, and (iv) what type of micro-forces can result in mechanical interactions on the boundary of type $\overset{(1)}{\tau} \cdot \boldsymbol{n}$?

Even if strain-gradient or generalized models are frequently evoked for explaining different effects in material science (e.g., size effects, regularization of stress concentration at the tip of fractures, ...), they have not the same status as classical material. They remain an abstract concept. Indeed, if in an experimenter desires to use, say an isotropic material with Young's modulus 200 GPa $= 2\ 10^{11}\mathrm{Nm}^{-2}$ he can

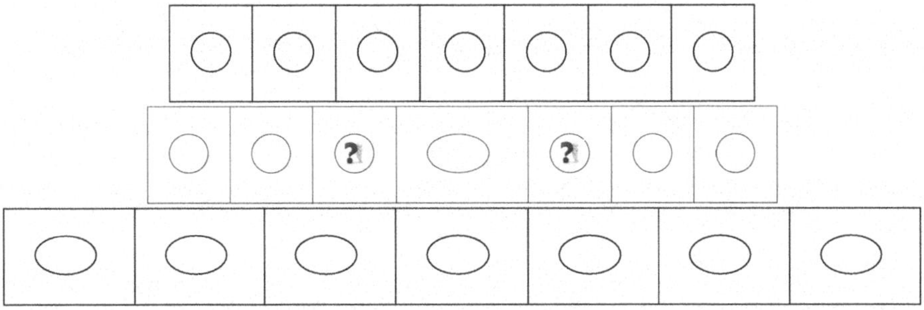

Fig. 4.1 Propagation of width. How to pass the information from one cell to its neighbors?

choose some steel. If he prefers a Poisson's ratio of 0.2, he may use some glass. If non-isotropic material is needed, wood- or fiber-reinforced composites can be used. Finding a material with negative Poisson's ratio becomes difficult: As they are scarce in nature, one could try to use a highly contrasted micro-architectured meta-material (see Milton 1992; Milton and Cherkaev 1995). But what about using a material with strain-gradient stiffness of, let say, $10^9 \mathrm{Nm}^{-4}$? There is no such material at our disposal, even approximately. A current challenge is to make at least one material with an indisputable strain-gradient behavior.

The idea for making such a material, both from the theoretical and hopefully from the practical point of view, is to conceive a meta-material. We will deal with architectured periodic structures and use homogenization theory. Homogenization of periodic elastic media is now a well-founded theory, and well-known results state that the effective behavior is still a classical elastic behavior. So, we need to get out of the standard assumptions leading to these results. We do that by assuming a very high contrast between the materials the periodic structure is made of. But that is not enough. If one desires to obtain a significant strain-gradient effect, one must conceive a structure such that the strain of a cell is influenced by the strain of its neighbors.

Let us illustrate this issue in a completely different framework: Consider a row of biologic cells (Fig. 4.1). Assume that one of the cells increases its width. How can the next cells be informed of this change of size in order to do the same by mimicry ? The simple fact that they are pushed apart is not significant: Without extra information, they cannot make the difference between a width increase of their neighbor and a global displacement of the row. The answer is clear: The growing cell must associate with its change of width some information along its boundary in order to establish a dialogue. For a biologic cell, it could be an electrical or chemical signal. In a purely mechanical framework, a mechanism must link the change of width to some special motion on the boundary. The pantographic idea is one of the simplest ways for doing so (Fig. 4.2): An extension of a cell of a pantograph is transformed in a shrinking of a pair of points (C, D) on the boundary of the cell and reciprocally.

As far as generalized continua are concerned, one has to conceive a cell with an internal degree of freedom, that is an internal motion, for instance the rotation of some substructure in a Cosserat-type model, then to allow for a weak coupling between

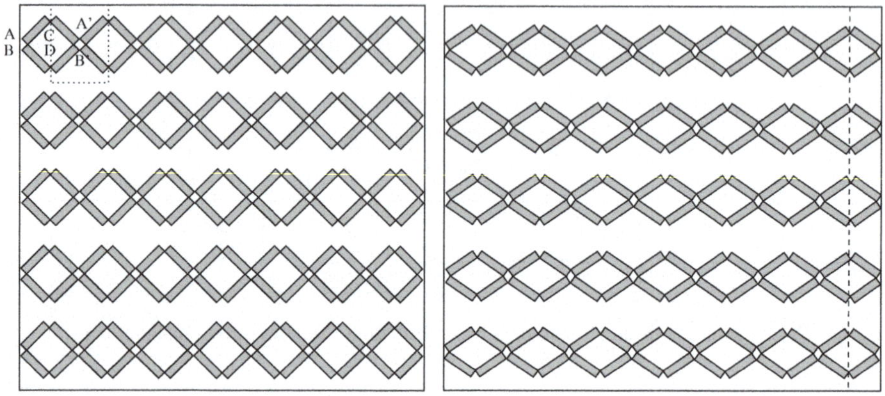

Fig. 4.2 Idea for simulating a strain-gradient material. A "pantographic" cell, where the gray part is much stiffer than the surrounding matrix, transforms its horizontal dilatation in a vertical shrinking between points C and D visible on its lateral boundary and reciprocally

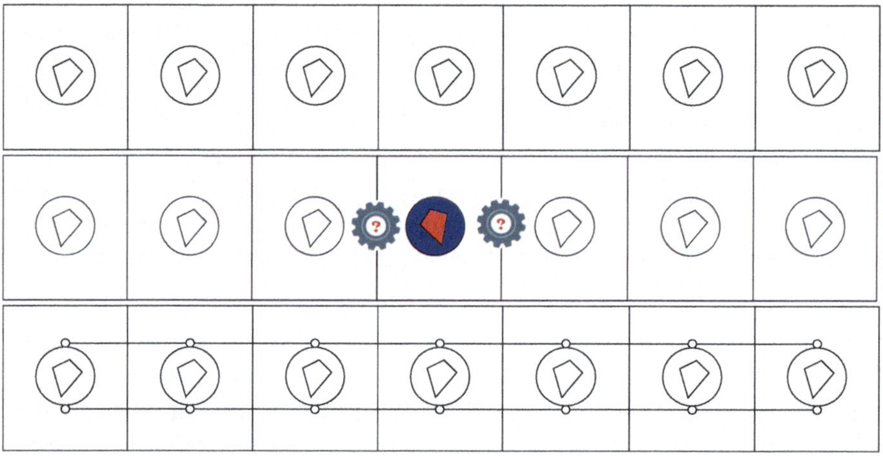

Fig. 4.3 Propagation of an internal rotation. To pass the information from a cell to the next one, a possibility is to add tie rods

this degree of freedom and the displacement field and finally to find out a way for coupling the internal motion of two neighboring cells. This last step, illustrated in Fig. 4.3, is often forgotten when explaining Cosserat effects (for instance, in granular materials).

The examples provided in Figs. 4.2 and 4.4 illustrate also the way microscopic forces give rise to the macroscopic mechanical interactions associated with strain-gradient or Cosserat continuum. In the pantographic structure described in Fig. 4.2, a pair of vertical opposite forces applied in the vicinity of points A and B will lead to a boundary action $\overset{(2)}{T} \cdot n$ dual to the horizontal expansion (a double force in the

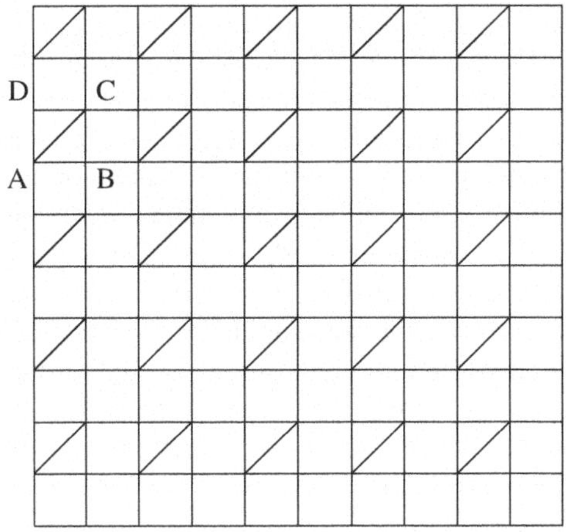

Fig. 4.4 Idea for simulating a Cosserat model: a square frame lattice with only one square containing an internal diagonal bar AC for every four squares. The square with the diagonal bar behaves essentially like a rigid body which has its own rotation. Coupling this rotation with the global rotation is ensured by the welding and the flexion stiffness of the vertical and horizontal bars. The coupling of the rotations in two neighboring cells is ensured by the extensional stiffness of the vertical and horizontal bars

nomenclature of Germain 1973a). Note how counterintuitive is the fact that a pair of microscopic vertical forces can be associated with a horizontal dilatation. Note also that a pair of horizontal opposite forces applied, for instance, in the vicinity of points A and A' would lead to the same boundary action. In the structure described in Fig. 4.4, pairs of microscopic forces acting like torques at points A and D lead to a macroscopic boundary action $\overset{(1)}{T} \cdot n$ corresponding to a surface density of torques.

The Homogenization Framework

We limit ourselves to static linear elasticity. The structure is contained in the domain $\mathcal{B}$, more precisely in a periodic subset $\mathcal{B}^\ell$ of $\mathcal{B}$. From the homogenization point of view, the macroscopic material is made of elastic material and void. We are dealing with an infinitely highly contrasted material, and we are therefore out of the framework of classical homogenization. Note that the term "architectured material" is more adapted to describe this situation than the term "composite material."

The global elastic energy of the material is $\mathcal{W}^\ell(\boldsymbol{u}) = \int_{\mathcal{B}^\ell} w^\ell(E_{\boldsymbol{u}})\,\mathrm{d}V$ with

$$w^\ell(E_u) = \frac{1}{2}(C^\ell \cdot\cdot E_u) \cdot\cdot E_u,$$

C^ℓ being a given positive-definite fourth-order isotropic stiffness tensor. The periodic cell $\ell\mathcal{Y}$ has characteristic size ℓ, and $\mathcal{Y}$ is the rescaled prototype cell.

The external bulk force is assumed to be concentrated on $\mathcal{B}^\ell$ which is a necessary condition for the existence of an equilibrium solution. As the total energy is convex, the system of equilibrium equations is equivalent to the minimization of the energy. We assume that ℓ is small compared to the domain characteristic length L. Thus, we consider a family of problems indexed by ℓ, called *microscopic problems*, and let ℓ tend to zero. Note that, even if $\mathcal{B}^\ell$ is a periodic domain, the corresponding *microscopic equilibrium solution* $\boldsymbol{u}^\ell$ need not to be periodic but presents oscillations with a length scale of order ℓ. We are interested in the limit $\boldsymbol{u}$ of the sequence $(\boldsymbol{u}^\ell)$ or at least of some subsequence of $(\boldsymbol{u}^\ell)$, and we hope that such a limit minimizes some energy, called *macroscopic (or effective or homogenized) energy* of the architectured material. As we are not interested in a particular bulk force, we desire to identify an elastic energy $\mathcal{W}$ independent of it.

Having accepted voids in the considered structures is not enough to get out of the scope of classical homogenization results which gives only classical elastic behavior. We have explained in the previous section that particular geometries have to be designed in order to "pass the desired information" from one cell to the next one. One can see, when considering the examples of Figs. 4.2 and 4.4, that the structure inside the cell must act like a mechanism and thus must contain *hinges*. Perfect hinges cannot be achieved, but they can be approached by introducing thin necks or more generally thin substructures. Hence, the geometry we have to consider, even in the rescaled cell $\mathcal{Y}$, must still depend on ℓ. Moreover, as the structure contains thin parts, the stiffness of the material must be large enough, when compared to the external applied loads, for the structure to be able to resist to them.

Generally, strain-gradient or micromorphic models contain at least one intrinsic length L^{int}. For instance, if a strain-gradient energy density contains, in addition to a classical strain energy with Young's modulus Y, a term $\chi(\partial_1(E_{\boldsymbol{u}})_{11})^2$ associated with the variation of dilatation in direction x_1, then the quantity $L^{int} := \sqrt{Y^{-1}\chi}$ is an intrinsic length. This length could be measured an expansion in direction x_1 on a part of the sample and measuring the distance for this expansion to be damped in the free part of the sample. Note that such a strain-gradient model forbids any strain discontinuity and regularizes it over a transition zone of thickness of order L^{int}.

A majority of authors who get strain-gradient effects through a homogenization process obtain only terms of order ℓ^2 which appear then to be corrective terms of a classical limit energy. In this way, they get "macroscopic" models which involve an intrinsic length of the same order as the microscopic one: ℓ. Our goal is very different: We want to describe a strain-gradient model with a real macroscopic intrinsic length. Indeed, there are two reasons for preferring a macroscopic description rather than a microscopic one. The first one is the prohibitive cost of a full simulation at the microscopic scale. The second one, less often invoked but maybe more important, is the fact that the structure of the material is not known at the microscopic scale with such a precision nor are external actions, especially on the boundary. When dealing with an intrinsic length of the same order as the microscopic cell size, these two concerns remain.

Even if we do not indent to give in this chapter the details of the proof of the homogenization result that we will state in section "A General Result for Structures Based on a Periodic Graph", let us quickly describe the principal mathematical tools that are used for identifying the effective behavior of architectured materials.

Mathematical Tools: Γ-convergence

The mathematical notion of Γ-convergence (see Braides 2002; Dal Maso 1993) precisely matches the intuitive notion of convergence of constitutive laws. Indeed, applied external forces are irrelevant and only the elastic energy $\mathcal{W}^\ell$ is considered.

Proving a useful Γ-convergence result is proving three items:

- First one has to prove that the structure remains stable when ℓ tends to zero: The equilibrium displacement field does not blow up. More precisely, for any M, any sequence $(\boldsymbol{u}^\ell)$ satisfying $\mathcal{W}^\ell(\boldsymbol{u}^\ell) < M$ contains a converging subsequence.
- Then, one has to prove that the expected effective energy $\mathcal{W}(\boldsymbol{u})$ associated with any macroscopic displacement field $\boldsymbol{u}$ is the smallest possible energy of all microscopic displacement fields $\boldsymbol{u}^\ell$ close to $\boldsymbol{u}$. This needs to prove that
  - $\mathcal{W}(\boldsymbol{u})$ is not too large: For any sequence $\boldsymbol{u}^\ell$ converging to $\boldsymbol{u}$ and such that $\mathcal{W}^\ell(\boldsymbol{u}^\ell)$ converges, $\lim \mathcal{W}^\ell(\boldsymbol{u}^\ell) \geq \mathcal{W}(\boldsymbol{u})$.
  - $\mathcal{W}(\boldsymbol{u})$ is not too small: For any $\boldsymbol{u}$, there exists a sequence $\boldsymbol{u}^\ell$ converging to $\boldsymbol{u}$ such that $\lim \mathcal{W}^\ell(\boldsymbol{u}^\ell) = \mathcal{W}(\boldsymbol{u})$.

As an exercise, we let the reader check that the following three one-dimensional (here the variable u is a real number) sequences of potentials

$$\mathcal{W}^\ell(u) = min\left(u^2, \left|\frac{u-\ell}{\ell}\right| - 1\right);$$

$$\mathcal{W}^\ell(u) = u^2 + \cos\left(\frac{u}{\ell}\right);$$

$$\mathcal{W}^\ell(u) = \begin{cases} u^2 & \text{if } u < 0 \\ u^2 + 1 & \text{if } u \geq 0 \end{cases};$$

Γ-converge, as ℓ tends to zero, respectively, to

$$\mathcal{W}(u) = \begin{cases} u^2 & \text{if } u \neq 0 \\ -1 & \text{if } u = 0 \end{cases};$$

$$\mathcal{W}(u) = u^2 - 1;$$

$$\mathcal{W}(u) = \begin{cases} u^2 & \text{if } u \leq 0 \\ u^2 + 1 & \text{if } u > 0 \end{cases};$$

and convince himself that the energies $\mathcal{W}$ thus obtained are the right ones for describing the limits of the equilibrium states, and that even in the presence of any external force.

Let us remark that, as usual, the definition of the limit does not provide the limit itself but just a way of proving that a guessed limit is the right one.

A first fundamental property of Γ-convergence is the fact the convergence result remains valid if any continuous linear functional is added. Hence, a result obtained without considering any external forces can be applied in the presence of such forces. A second property is the fact that the three previous items imply the convergence of (sub)sequence of equilibriums for $\mathcal{W}^\ell$ toward an equilibrium for $\mathcal{W}$.

One may remark that, when u belongs to a functional space, the choice of topology that one uses while proving the three items seems crucial. This should be a concern as it would not be "physical" to have different macroscopic behaviors depending on the mathematical choice of a topological framework. Fortunately, another property of Γ-convergence states that all topologies ensuring the previous three items lead to the same Γ-limit.

Mathematical Tools: Double-Scale Convergence

For proving a Γ-limit result, one needs to have an accurate description of any minimizing sequence u^ℓ. The functions u^ℓ have an oscillating behavior at the scale of the periodic cell $\ell\mathcal{Y}$. Double-scale asymptotic expansion is a powerful tool for capturing this phenomenon. Following Nguetseng (1989) and Allaire (1992), we say that u^ℓ double-scale converges to u (and denote $u^\ell \rightharpoonup\rightharpoonup u$) if, for any y-periodic test function ψ,

$$\int_{\mathcal{B}} u^\ell(x) \cdot \psi\left(x, \frac{x}{\ell}\right) \mathrm{d}V \to \int_{\mathcal{B}} \int_{\mathcal{Y}} u(x, y) \cdot \psi(x, y) \, \mathrm{d}V_y \, \mathrm{d}V_x$$

As an exercise, we let the reader check that the sequences of functions

$$u^\ell(x) := x^2 + \sin(x/\ell);$$
$$u^\ell(x) := x^2 + \sin(x/\ell^2)$$

double-scale converge, respectively, to

$$u(x, y) := x^2 + \sin(y);$$
$$u^\ell(x, y) := x^2$$

We see on these examples that this notion captures the oscillations are scale ℓ but discards oscillations at a smaller scale. A first property of double-scale convergence states that any sequence bounded in L^2 possesses some subsequence which double-scale converges. A second useful property is the fact that, if $u^\ell \rightharpoonup\rightharpoonup u$, then, for any convex function w,

$$\liminf \int_{\mathcal{B}} w(u^\ell(x))\,\mathrm{d}V \geq \int_{\mathcal{B}} \int_Y w(u(x,y))\,\mathrm{d}V_y\,\mathrm{d}V_x.$$

A General Result for Structures Based on a Periodic Graph

We will now state a general homogenization theorem for structures based on a periodic graph. The goal is to determine the effective behavior of structures of the type represented in Fig. 4.5.

Even if more general cases can be treated, we limit ourselves in this chapter to structures which are cylinders based on a 2D periodic graph. They are subjected to forces or imposed displacements such that we can work in the framework of plane-strain linear elasticity. The problem becomes a 2D linear elasticity problem set on a $\ell\mathcal{Y}$-periodic domain $\mathcal{B}^\ell$. We moreover assume that the structure is made by slender substructures: A cell contains a finite number K of rigid nodes which are linked (or not) by thin bars (or "walls" if one refers to the structure of Fig. 4.5). Similar walls may also link some nodes of the cell to nodes of the neighboring cells. We emphasize that there is no loss of generality by assuming that a cell is linked only to its closest neighboring cells. Indeed, one is allowed to choose a larger period (made of several elementary cells) for this assumption to become valid.

It is clear that the slenderness of the walls makes the structure weak: With given Lamé coefficients (λ, μ), it can only resist to weak external forces. An equivalent

Fig. 4.5 A 3D-printed structure based on a periodic graph

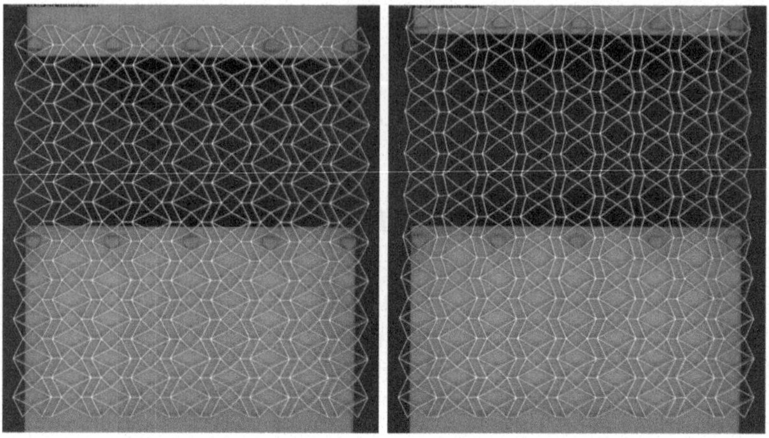

Fig. 4.6 A traction experiment for the structure of Fig. 4.5. The structure is fixed along its middle line and pulled up along its top boundary

way of thinking, that we prefer, is to say that the structure can resist to external forces of order one only if the Lamé coefficients are large enough. How large ? This depends on the slenderness of the walls but also on the geometry of the structure. Let L be the macroscopic size of the sample and ℓ the very small cell size. We will assume that the thickness $\beta\ell^2/L$ of the walls is of order ℓ^2/L. Indeed, this scaling is critical: Letting in a second step β tend to 0 or $+\infty$ gives the effective behavior when the thickness of the walls has a different order of magnitude. We will also focus on structures which have the specificity of being able to endow a global motion without extending any of the slender walls but only by bending them. Hence, the order of magnitude of the global stiffness of the structure must be estimated from the bending stiffness of the substructures: The Lamé coefficients must be of order $(\ell/L)^{-3}$.

We assume, for the sake of simplicity, that the structure is made of an isotropic elastic material. The elastic energy associated with a displacement field $\boldsymbol{u}$ reads

$$\mathscr{W}^\ell(\boldsymbol{u}) := \int_{\mathcal{B}^\ell} \left(\mu_0 \left(\frac{L}{\ell}\right)^3 \|E_{\boldsymbol{u}}\|^2 + \frac{\lambda_0}{2}\left(\frac{L}{\ell}\right)^3 (\mathrm{tr}(E_{\boldsymbol{u}}))^2 \right) \mathrm{d}V$$

A first step for computing the Γ-limit, as ℓ tends to zero, of this sequence of energies is estimating the energy of each substructure. This is a rather technical step which needs to remember one's structural mechanics class. Each node behaves almost like a small rigid body, and each slender bar (or wall) which links two nodes possesses an energy made of two parts: an extensional energy related to the relative displacement of the two nodes in the direction of the bar and a bending energy related to the differences between the rotations of the two nodes and the global rotation of the bar. Even if the bending energy is much smaller than the extensional energy, it plays an essential role in the effective behavior. These estimations transform the initial continuum mechanic problem into a discrete problem where the continuous elastic energy is replaced by non-local interactions between nodes. It can be shown (see Abdoul-Anziz and Seppecher 2018a) that this discrete problem actually leads to the same Γ-limit as the initial one.

Next section is devoted to the description of this discrete problem.

The Discrete Problem

For describing precisely the discrete problem, we need to fix the geometry by fixing a prototype cell $\mathcal{Y}$ in which lie K nodes $\boldsymbol{y}^k$, $k \in \{1, \ldots, K\}$ and two independent periodicity vectors $\boldsymbol{t}^1$ and $\boldsymbol{t}^2$. Hence, the nodes of the structure are the points[3] $\ell(i\boldsymbol{t}^1 + j\boldsymbol{t}^2 + \boldsymbol{y}^k)$ for all $k \in \{1, \ldots, K\}$ and all i and j in $\{0, \ldots, \ell^{-1} - 1\}$ (for the sake of simplicity, ℓ^{-1} is assumed to be integer). Once the position of the nodes is fixed, we can determine the position of the bars by choosing the pairs of nodes which are linked. A first matrix a_1 defines the bars inside a cell: $(a_1)_{s,r} \neq 0$ means that nodes $\ell(i\boldsymbol{t}^1 + j\boldsymbol{t}^2 + \boldsymbol{y}^s)$ and $\ell(i\boldsymbol{t}^1 + j\boldsymbol{t}^2 + \boldsymbol{y}^r)$ are linked and the value of $(a_1)_{s,r}$ (simply denoted $a_{1,s,r}$) corresponds to the extensional stiffness of the bar which links them. A second matrix a_2 defines the bars which link some nodes of a cell with the nodes of the next cell following direction $\ell\boldsymbol{t}^1$: $(a_2)_{s,r} \neq 0$ means that nodes $\ell(i\boldsymbol{t}^1 + j\boldsymbol{t}^2 + \boldsymbol{y}^s)$ and $\ell((i+1)\boldsymbol{t}^1 + j\boldsymbol{t}^2 + \boldsymbol{y}^r)$ are linked. Similar matrices a_3, a_4, a_5 correspond to links with the next cell following, respectively, direction $\ell\boldsymbol{t}^2$, $\ell(\boldsymbol{t}^1 + \boldsymbol{t}^2)$ and $\ell(\boldsymbol{t}^1 - \boldsymbol{t}^2)$.[4]

To resume, the geometry is completely described by two vectors $\boldsymbol{t}^1, \boldsymbol{t}^2$, K points $\boldsymbol{y}^k$ and $5 \times K \times K$ quantities $a_{p,r,s}$.

For instance, the structure represented in Fig. 4.7 is defined by

$$K = 2, \quad \boldsymbol{y}_1 = (0,0), \quad \boldsymbol{y}_2 = (0,1), \quad \boldsymbol{t}^1 = (1,0), \quad \boldsymbol{t}^2 = (0,2),$$

$$a_1 = \begin{pmatrix} 0 & 1 \\ 0 & 0 \end{pmatrix}, a_2 = \begin{pmatrix} 1 & 1 \\ 0 & 1 \end{pmatrix}, a_3 = \begin{pmatrix} 0 & 0 \\ 1 & 0 \end{pmatrix}, a_4 = \begin{pmatrix} 0 & 0 \\ 0 & 0 \end{pmatrix}, a_5 = \begin{pmatrix} 0 & 0 \\ 0 & 0 \end{pmatrix}.$$

We will see later (in Sect. 4.5.5) that this structure leads to a couple-stress effective behavior. The reader can try to define in this way the different examples that we consider in this paper, and many others, and check these structures using the online tool:
http://seppecher.univ-tln.fr/homogenizer/truss.html.

When all nodes $\ell(i\boldsymbol{t}^1 + j\boldsymbol{t}^2 + \boldsymbol{y}^s)$ move with respective displacement $\boldsymbol{U}_{(i,j),s}$, then the presence of bars linking nodes leads to an extensional energy. In linear

---

[3] The way the nodes of the structure depend here on ℓ, while natural, could be questioned: In Abdoul-Anziz et al. (2019), a different dependence is proved to lead in the effective energy to a coupling of strain and strain-gradient terms.

[4] For $p \in \{1, \ldots, 5\}$, notation $(i,j)+p$ stands for the corresponding couple in $\{(i,j), (i+1,j), (i,j+1), (i+1,j+1), (i+1,j-1)\}$ and notation $\boldsymbol{p}$ stands for the corresponding vector in $\{0, \boldsymbol{t}^1, \boldsymbol{t}^2, \boldsymbol{t}^1 + \boldsymbol{t}^2, \boldsymbol{t}^1 - \boldsymbol{t}^2\}$.

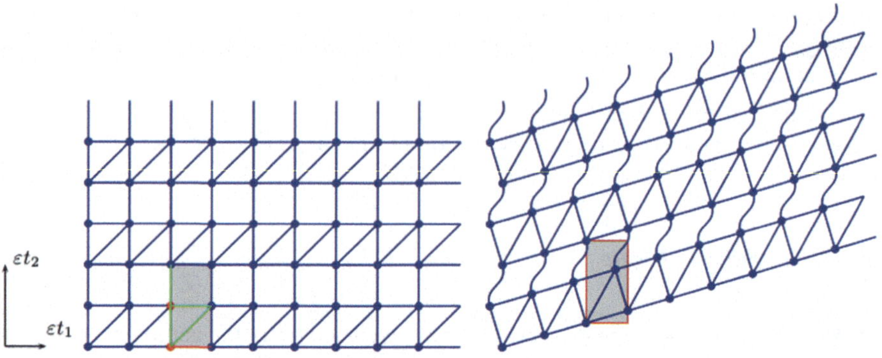

Fig. 4.7 Example of a periodic lattice. This structure has the specificity that it can undergo a global non-trivial deformation while the distance between every pair of linked nodes remains unchanged

elasticity, the extension between two nodes (I, s) and $(I + p, r)$ is measured by

$$\frac{\boldsymbol{U}_{I+p,s} - \boldsymbol{U}_{I,r}}{\ell} \cdot \boldsymbol{d}_{p,r,s}, \text{ where } \boldsymbol{d}_{p,r,s} := \frac{\boldsymbol{p} + \boldsymbol{y}^s - \boldsymbol{y}^r}{\|\boldsymbol{p} + \boldsymbol{y}^s - \boldsymbol{y}^r\|}$$

stands for the unit vector in the direction of the bar. The global extensional energy of this bar is proportional to the square of this quantity. All textbooks of structural mechanics give the proportionality coefficient in terms of the length and section area of the bar: Setting

$$a_{p,r,s} = \frac{8\mu_0(\mu_0 + \lambda_0)}{2\mu_0 + \lambda_0} \frac{\beta L^3}{\|\boldsymbol{p} + \boldsymbol{y}^s - \boldsymbol{y}^r\|},$$

the global extensional energy of the structure reads

$$\mathscr{W}_\ell^{ext}(\boldsymbol{U}) := \sum_{I,p,r,s} \frac{a_{p,r,s}}{2} \left(\frac{\boldsymbol{U}_{I+p,s} - \boldsymbol{U}_{I,r}}{\ell} \cdot \boldsymbol{d}_{p,r,s} \right)^2$$

where the sum is carried out over all $I = (i, j)$, $p \in \{1, \dots, 5\}$ and $(r, s) \in \{1, \dots, K\}^2$ with the convention that $a_{p,r,s} = 0$ as soon as no bar links nodes (I, r) and $(I + p, s)$.

The effect of bending stiffness is a bit more difficult to compute. One has to take into account not only the displacements of all nodes $\ell(i\boldsymbol{t}^1 + j\boldsymbol{t}^2 + \boldsymbol{y}^s)$ but also their rotation $\alpha_{(i,j),s}$. The global bending energy of a bar depends on the difference of rotation between the two endpoints and on the difference between these rotations and the global rotation

$$r_{p,r,s} := \frac{\boldsymbol{U}_{I+p,s} - \boldsymbol{U}_{I,r}}{\ell\|\boldsymbol{p} + \boldsymbol{y}^s - \boldsymbol{y}^r\|} \cdot \boldsymbol{d}^\perp_{p,r,s}$$

The global bending energy of the structure reads

$$\mathscr{W}_\ell^{flex}(\boldsymbol{U},\alpha) := \left(\frac{\ell}{L}\right)^2 \sum_{I,p,r,s} \frac{a_{p,r,s}}{2}\frac{\beta^2}{6}\left(3(\alpha_{I+p,s}+\alpha_{I,r}-2r_{p,r,s})^2+(\alpha_{I,r}-\alpha_{I+p,s})^2\right).$$

The global energy of the structure is the sum of the extensional and bending energies. As we desire to describe the macroscopic effective behavior in terms of the limit displacement field only, we deal with the energy

$$\mathscr{W}_\ell(U) := \inf_\alpha\{\mathscr{W}_\ell^{ext}(\boldsymbol{U}) + \mathscr{W}_\ell^{flex}(\boldsymbol{U},\alpha)\}. \tag{4.13}$$

The Homogenization Result for Structures Based on Periodic Graphs

The double-scale convergence is suitable for treating classical homogenization problems. It must be adapted to deal with the sequences of displacements or rotations which are defined only at points $\boldsymbol{y}_s$ of each periodic cell I but however converge in some sense to continuous functions. In that case, it is natural to say that $u^\ell(I,s)$ double-scale converges to $u(x,y_s)$ if for any test function ψ, $\frac{\ell^2}{L^2}\sum_I\sum_s u^\ell_{I,s}\psi(\ell(it^1+jt^2),\boldsymbol{y}_s) \to \frac{1}{|\mathcal{B}|}\int_{\mathcal{B}}\sum_s u^\ell(\boldsymbol{x},\boldsymbol{y}_s)\psi(x,\boldsymbol{y}_s)$ or, in a much more simple way, to say that $u^\ell_{I,s}$ double-scale converges to $u_s(x)$ if

$$\forall s,\ \forall\psi,\ \frac{\ell^2}{L^2}\sum_I\sum_s u^\ell_{I,s}\psi(\ell(it^1+jt^2)) \to \frac{1}{|\mathcal{B}|}\int_{\mathcal{B}}\sum_s u_s(\boldsymbol{x})\psi(\boldsymbol{x}).$$

Once the notion of convergence is fixed, we can identify the Γ-limit of the sequence of energies $\mathscr{W}_\ell$. The two first points of Γ-convergence are rather technical and out of the scope of the present chapter. But, for identifying the Γ-limit, it is enough to consider smooth functions $\boldsymbol{u}$, $\boldsymbol{v}_s$, $\boldsymbol{w}_s$, α_s and sequences of the type

$$\boldsymbol{U}^\ell_{I,s} := \boldsymbol{u}(\boldsymbol{y}^\ell_I) + \ell\boldsymbol{v}_s(\boldsymbol{y}^\ell_I) + \ell^2\boldsymbol{w}_s(\boldsymbol{y}^\ell_I),\quad \alpha^\ell_{I,s} := \alpha_s(\boldsymbol{y}^\ell_I), \tag{4.14}$$

and to compute the limit energy of such sequences which happens to be the researched Γ-limit (moreover, such sequences can be used as recovery sequences in the sense of point (iii) of Γ-convergence).

Plugging developments (4.14) inside the expression (4.13) of the energy is straightforward: The leading term is $\left(\frac{\ell}{L}\right)^{-2}$ times a quantity tending to $\bar{\mathscr{W}}^{ext}(\boldsymbol{v},\eta_{\boldsymbol{u}})$ where

$$\bar{\mathscr{W}}^{ext}(\boldsymbol{v},\eta) := \int_{\mathcal{B}} \sum_{p,s,s'} \left(\frac{a_{p,s,s'}}{2} \left(\left(\boldsymbol{v}_{s'}(x) - \boldsymbol{v}_s(x) + \eta_{p,s'}(x) \right) \cdot \boldsymbol{d}_{p,s,s'} \right)^2 \right) \mathrm{d}V \quad (4.15)$$

and $(\eta_{\boldsymbol{u}})_{p,s} := \nabla \boldsymbol{u} \cdot \boldsymbol{p}$. Hence, any sequence with bounded energy (and these are the only interesting sequences) must satisfy

$$\bar{\mathscr{W}}^{ext}(\boldsymbol{v},\eta_{\boldsymbol{u}}) = 0.$$

The minimizers $\boldsymbol{v}$ of $\bar{\mathscr{W}}^{ext}(\boldsymbol{v},\eta_{\boldsymbol{u}})$ depend linearly on $\eta_{\boldsymbol{u}}$ up to an element $\tilde{\boldsymbol{v}}$ of the kernel K of this energy: Let us introduce a linear operator L such that $\boldsymbol{v} = L(\eta_{\boldsymbol{u}}) + \tilde{\boldsymbol{v}}$, $\tilde{\boldsymbol{v}} \in K$. Indeed, the energy is degenerate: Its kernel contains at least the constant fields $\boldsymbol{v}$ but the most interesting cases are when the kernel is larger. The fact that $\bar{\mathscr{W}}^{ext}(\boldsymbol{v},\eta_{\boldsymbol{u}}) = 0$ has two important consequences:

$$\boldsymbol{v} = L(\eta_u) + \tilde{\boldsymbol{v}} \text{ with } \tilde{\boldsymbol{v}} \in K \quad \text{and} \quad \bar{\mathscr{W}}^{ext}(L(\eta_u),\eta_u) = 0 \qquad (4.16)$$

the second equation shows that, in general, the effective behavior of the considered structures is constrained: The strain tensor must belong to some subspace. For instance, the homogenized material may be incompressible or inextensible in some direction. This should not seem too strange to the reader: Incompressible models are common in material science. For instance, rubber is said to be incompressible. Actually, rubber is not really incompressible but the point is that it is very soft with respect to shear. Thus, one generally acts on it with gentle forces which are not able to significantly modify its volume. A similar phenomenon arises here. The structures we consider are conceived in such a way that they allow for global deformations which do not require the extension of any bar (see Fig. 4.7). Applied forces have the needed order of magnitude for activating these modes of deformation and are too weak for activating the other modes of deformation which require the extension of the bars. These last deformation modes appear then to be "forbidden" by the model.

Once the constraints (4.16) are respected, the leading order term in the expression of the energy disappears and we can compute the limit energy. The result, which needs some technical mathematical work, can be written $\bar{\mathscr{W}}^{ext}(\boldsymbol{w},\xi_{\boldsymbol{u},\boldsymbol{v}}) + \bar{\mathscr{W}}^{flex}(\boldsymbol{v},\eta_{\boldsymbol{u}},\alpha)$ where $(\xi_{\boldsymbol{u},\boldsymbol{v}})_{p,s} := \nabla \boldsymbol{v}_s \cdot \boldsymbol{p} + \frac{1}{2}\nabla\nabla\boldsymbol{u}\cdot\boldsymbol{p}\cdot\boldsymbol{p}$ and $\bar{\mathscr{W}}^{flex}(\boldsymbol{v},\eta,\alpha)$ stands for

$$\int_{\mathcal{B}} \sum_{p,s,s'} \frac{a_{p,r,s}}{2} \frac{\beta^2 L^2}{6} \left(3(\alpha_s(x) + \alpha_r(x) - 2\bar{r}_{p,r,s}(x))^2 + (\alpha_r(x) - \alpha_s(x))^2 \right),$$

with $\bar{r}_{p,r,s}(x) := \dfrac{\boldsymbol{v}_s(x) - \boldsymbol{v}_r(x) + \eta_{p,s}(x)}{\|\boldsymbol{p} + \boldsymbol{y}^s - \boldsymbol{y}^r\|} \cdot \boldsymbol{d}^{\perp}_{p,r,s}$. To resume, the limit energy, in terms of the displacement field $\boldsymbol{u}$ only, reads

$$\boxed{\mathscr{W}(\boldsymbol{u}) := \inf_{\boldsymbol{w},\boldsymbol{v},\alpha} \left\{ \bar{\mathscr{W}}^{ext}(\boldsymbol{w},\xi_{\boldsymbol{u},\boldsymbol{v}}) + \bar{\mathscr{W}}^{flex}(\boldsymbol{v},\eta_{\boldsymbol{u}},\alpha);\ \bar{\mathscr{W}}^{ext}(\boldsymbol{v},\eta_{\boldsymbol{u}}) = 0 \right\}}$$

Three extra kinematic variables, namely $\boldsymbol{w}$, $\boldsymbol{v}$ and α, enter this expression. Eliminating $\boldsymbol{w}$ and α is straightforward. Minimizing with respect to these variables is a local problem (a "cell problem"): One has to solve a linear system in a low dimension ($3K$) space. Using constraint (4.16), $\boldsymbol{v}$ can be replaced by $\tilde{\boldsymbol{v}}$ and we end up with a limit energy of the type

$$\boxed{\mathscr{W}(\boldsymbol{u}) := \inf_{\tilde{\boldsymbol{v}}} \int_{\mathcal{B}} w^{ext}(\nabla\nabla\boldsymbol{u}, \nabla\tilde{\boldsymbol{v}}) + w^{flex}(\nabla\boldsymbol{u}, \tilde{\boldsymbol{v}})} \tag{4.17}$$

where w^{ext} and w^{flex} are explicit nonnegative quadratic forms.

We recognize here a strain-gradient energy together with an extra kinematic variable. The effective behavior of the considered structures falls within the framework of "generalized continua." In some cases, the variable $\tilde{v}$ can still be eliminated and we get a pure strain-gradient model. The following examples will show that all cases are interesting.

Examples

For each of the following structures, we do not detail the way the structures are defined as the figures are sufficiently clear. Remind however that we are considering frame lattices. Bars are welded when they join. Even if we have represented nodes in our figures by small balls, there is no hinge there. We do neither detail the computation of the effective energy: One has simply to follow the procedure described in the previous section. We simply describe the homogenized behavior by giving the expression of the limit energy, possibly completed by the constraint that the strain has to satisfy. We let the reader write the stiffness matrices associated with the energy. Such matrices take more space and, in our opinion, are less intelligible. We do neither provide the exact values of the different coefficients which enter the expressions of the effective energies: Indeed, they depend on the energy unit and on the chosen thickness for the bars. Remind that our goal is to illustrate the different types of models that lattices can lead to.

Regular Triangle Lattice

We start by checking a trivial example. Let us recall that our limit model is adapted to forces we are only able to activate deformation modes which do not require a significant extension of the bars. It is well know that in a triangulated structure no such modes exist apart from the rigid motions. Indeed, for the regular triangle lattice represented in Fig. 4.8, constraint (4.16) reads $E_{11} = E_{22} = E_{12} = 0$: The strain tensor E must vanish. The effective model corresponds to an undeformable 2D continuum.

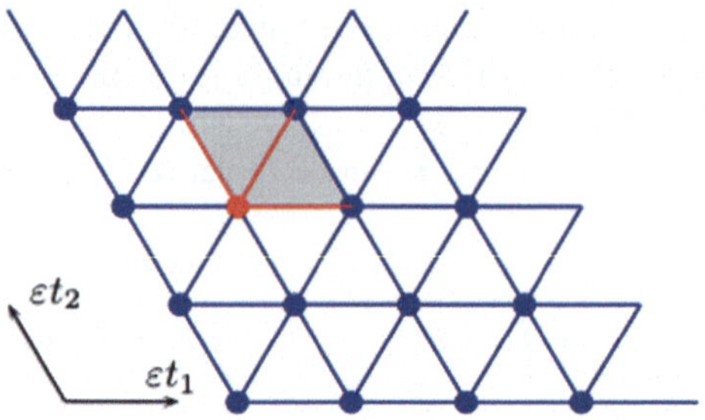

Fig. 4.8 The regular triangle lattice possesses no global motion (other than rigid body motions) which does not require the extension of the bars

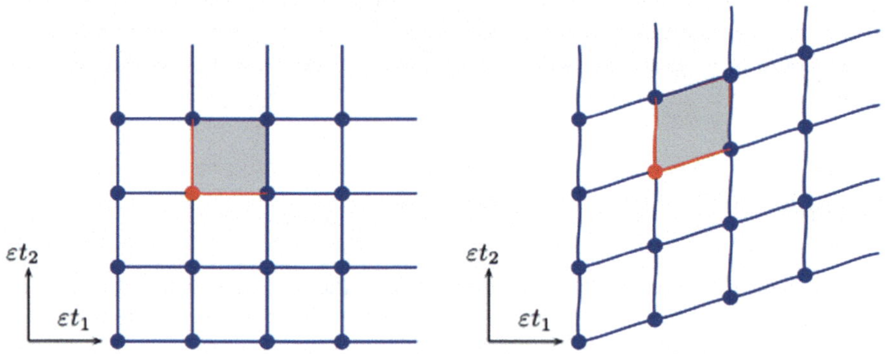

Fig. 4.9 The uniform shear of the square lattice can be performed without extension of any bar

Square Grid

For the square lattice represented in Fig. 4.9, constraint (4.16) reduces to $E_{11} = E_{22} = 0$. Hence, shear E_{12} remains free. One can see in Fig. 4.9 that such a shear can be performed without extending any bar (remind that we work in the framework of small displacement elasticity). However, it is clear that such a shear imposes the bending of the bars and thus needs some deformation energy. Indeed, our computation leads to the following effective energy:

$$\mathscr{W}(\boldsymbol{u}) = \int_{\mathcal{B}} \chi\,(E_{12})^2 \,\mathrm{d}V.$$

The effective behavior is the one of a "classical" 2D elastic material, but inextensible in the two orthogonal directions of the "fibers."

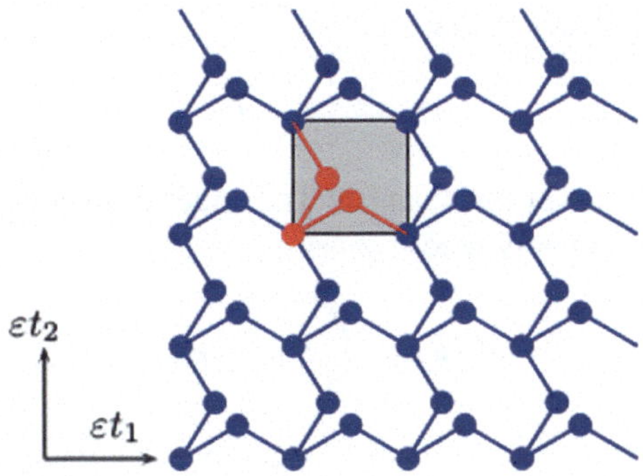

Fig. 4.10 A lattice with a non-constrained effective energy

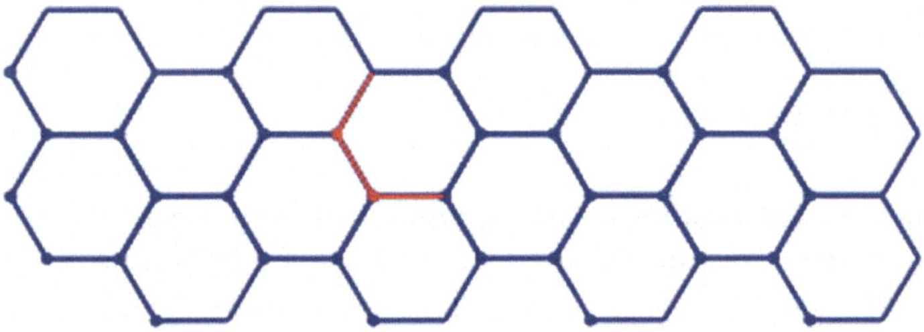

Fig. 4.11 Honeycomb structure

Square Grid Without Any Constraint

This example, described by Fig. 4.10, is provided in order to show that non-constrained materials can also be attained. We have replaced in the previous square grid the vertical and horizontal bars by zigzags.

Indeed, constraint (4.16) disappears in that case and we obtain the following effective energy:

$$\mathscr{W}(\boldsymbol{u}) = \int_{\mathcal{B}} \xi\,(E_{11})^2 + \xi\,(E_{22})^2 + \lambda\,(E_{12})^2 \,\mathrm{d}V.$$

This lattice, once homogenized, behaves like a classical non-isotropic elastic material.

Honeycomb Structure

The honeycomb lattice (see Fig. 4.11) is widely used because of its special mechanical properties. Many authors even think it shows strain-gradient effects.

Constraint (4.16) reduces in that case to

$$E_{11} + E_{22} = 0.$$

The trace of the strain tensor must vanish: The effective material is incompressible. Moreover, the effective energy reads

$$\mathscr{W}(\boldsymbol{u}) = \int_{\mathcal{B}} \chi \|E\|^2 \,\mathrm{d}V$$

The effective model is the one of a classical incompressible isotropic elastic material: No generalized continuum nor strain-gradient effects are present in the effective model.

Couple-Stress Structure

This example, already illustrated in Fig. 4.7, is the first one which goes out of classical elasticity and even of classical Cauchy theory of continuous media. The idea is to reconsider the square lattice by adding diagonal bars. If one adds such diagonal bars in every square, then the lattice would again be a triangular lattice and the effective material would be undeformable. That is why we add diagonal bars in one horizontal row of squares over two. Doing so, we still allow for a global shear without the extension of any bar. What is new is the fact that only *a uniform shear* is now possible without extending any bar. Like in the square lattice, shear needs the bending of the bars and this leads to a strain energy. But also a variation of the shear needs some small extension of the bars, and this leads to a strain-gradient energy. The needed extensions are small (of order ℓ/L), but this is compensated by the huge extensional stiffness of the bars. And indeed, applying the procedure we have previously described we get the effective energy

$$\mathscr{W}(\boldsymbol{u}) = \int_{\mathcal{B}} \chi \left(\frac{\partial^2 u_2}{\partial x_1^2}\right)^2 + \xi\,(E_{12})^2 \,\mathrm{d}V$$

under the same constraint as the square grid, namely $E_{11} = E_{22} = 0$. It is clear that the model enters the class of strain-gradient models. Actually, it enters the smaller class of couple-stress models, that is, models whose strain-gradient energy involves only the gradient of the curl of the displacement field. Indeed, owing to the constraint $E_{11} = 0$ we know that $\frac{\partial^2 u_1}{\partial x_2 \partial x_1} = \frac{\partial E_{11}}{\partial x_2} = 0$ and we can rewrite the energy under the form

$$\mathscr{W}(\boldsymbol{u}) = \int_{\mathcal{B}} \chi \left(\frac{\partial}{\partial x_1}\left(\frac{\partial u_2}{\partial x_1} - \frac{\partial u_1}{\partial x_2}\right)^2 + \xi\,(E_{12})\right)^2 \mathrm{d}V.$$

In a couple-stress model, the extra boundary conditions (4.12) correspond to a surface density of torques and thus are easier to understand than for general strain-gradient model. The lattice here gives a microscopic interpretation of this density of torques: It is easy to understand that applying opposite horizontal forces on the nodes lying on the right-hand boundary of the domain is applying torques on each horizontal "Warren beam" made by a horizontal row of squares with diagonals and that applying such torques tends to turn each of beam and thus to force a global shear of the structure.

It is important to remark that there is no natural mathematical way for associating a surface density of torques to a given distribution at the microscopic level of such horizontal boundary forces: Considering different pairs of successive nodes, one could find a positive or a negative density of torques at the macroscopic level. Only the mechanical intuition or the precise study of the Γ-limit, including boundary forces, can tell whether a positive or negative density of torques is applied to the couple-stress material.

Pantographic Lattice

We now go back to the printed structure represented in Figs. 4.5, 4.6 and 4.12. It is able to endow a shear without the extension of any bar, but the real specificity of this structure is its ability of being extended in the horizontal direction without needing the extension of any bar. More important is the fact that only a uniform global extension can be performed without the extension of any bar and the structure has been designed for that aim.

In that case, constraints (4.16) reduce to $E_{22} = 0$. Even if it is not obvious when considering the figure, the structure is inextensible in the vertical direction. Actually,

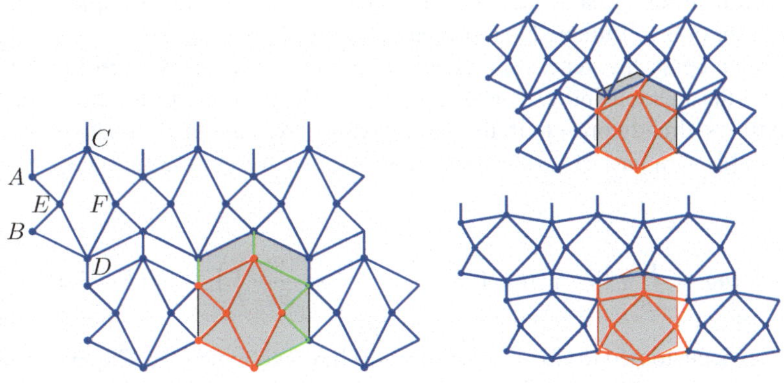

Fig. 4.12 A lattice leading to a general strain-gradient effective behavior and its two modes of deformation which do not require the extension of any bar (the required bending is not represented)

it has also been conceived in such a way that horizontal extension does not lead to vertical shrinking (zero Poisson effect). Then, we obtain an effective energy of the following type:

$$\mathscr{W}(\boldsymbol{u}) = \int_{\mathcal{B}} \xi\,(E_{12})^2 + \eta\,(E_{11})^2 + \chi\left(2\frac{\partial E_{12}}{\partial x_1} - \frac{\partial E_{11}}{\partial x_2}\right)^2 + \kappa\left(\frac{\partial E_{12}}{\partial x_1} - \kappa\frac{\partial E_{11}}{\partial x_2}\right)^2 + \chi\left(\frac{\partial E_{11}}{\partial x_1}\right)^2 \mathrm{d}V.$$

The fact that this model enters the class of strain-gradient models is obvious. The presence of the last addend prevents it to be a couple-stress model. Hence, the notion of double forces is needed to describe boundary conditions. We get here a microscopic interpretation of double forces: Assume that you are applying an horizontal force $-f\boldsymbol{e}^1$ on each node of type E along the left-hand boundary of the domain and the opposite force $f\boldsymbol{e}^1$ on each corresponding node of type F, then you force the horizontal extension of each cell along this boundary, and the "pantographic" geometry of the lattice tends to propagate this extension in the whole domain. These pairs of opposite forces lead in the macroscopic model to a surface density of double forces. Note that, like we already mentioned, there exists another counterintuitive way for acting on the boundary with the same macroscopic effect: Indeed applying along the left-hand boundary, a vertical force $f\boldsymbol{e}^2$ on each node of type A and the opposite force $-f\boldsymbol{e}^2$ on nodes of type B also tend to extend horizontally the cells and thus lead to a similar surface density of double forces.

Structure Leading to a Cosserat 2D Continuum

In the examples we have treated up to now, the extra kinematic variable $\tilde{\boldsymbol{v}}$ can be eliminated from expression (4.17) of the effective energy. An example where this variable cannot be eliminated is represented in Fig. 4.13.

When we apply the general procedure to the lattice represented in Fig. 4.13, we again find the inextensibility constraints $E_{11} = E_{22} = 0$ due to the horizontal and vertical fibers already present in the square lattice. We cannot eliminate entirely the extra variable $\tilde{\boldsymbol{v}}$, a part of it denoted γ remains, and we get the following effective energy:

$$\mathscr{W}(\boldsymbol{u}) = \int_{\mathcal{B}} \chi\|\nabla\gamma\|^2 + \xi\left(\gamma - \frac{1}{2}\left(\frac{\partial u_2}{\partial x_1} - \frac{\partial u_1}{\partial x_2}\right)\right)^2 + \kappa\left(E_{12}\right)^2 \mathrm{d}V.$$

The reader accustomed to Cosserat media must have recognized it. The extra variable γ is generally associated with an internal rotation: We get a microscopic interpretation of it as indeed γ corresponds to the rotation of the square $ABDC$ which, at the leading order, follows a rigid motion. This rotation is coupled with the global motion of the

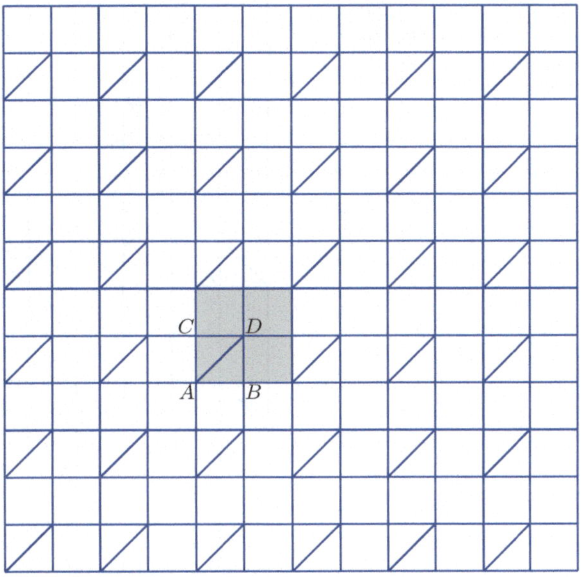

Fig. 4.13 A lattice leading to a generalized continuum

lattice as all bars have a bending stiffness and are welded together. This is the origin of the term $\xi\left(\gamma - \frac{1}{2}\left(\frac{\partial u_2}{\partial x_1} - \frac{\partial u_1}{\partial x_2}\right)\right)^2$. We also get an interpretation of the Cosserat term $\chi\left(\frac{\partial \gamma}{\partial x_1}\right)^2$ as the coupling between the rotation of a square $ABDC$ with the similar square in a neighbor cell is ensured by the extensional stiffness of the horizontal or vertical bars linking the two cells.

Cosserat and Second-Gradient Effect Together

It is remarkable that generalized continua effects arise at the same level as strain-gradient effects. This should close the controversy about which model of the two is more "natural." Are, like many authors claim, strain-gradient models nothing else than a limit case of generalized continua where the extra variable is constrained to coincide with the strain? Of course, but our results show that there is no hierarchy in the way the models arise through homogenization. The example represented in Fig. 4.14 shows that both effects can be present together in the effective model. We have conceived a lattice which mixes the properties of the two previous ones, and we get still the same constraints $E_{11} = E_{22} = 0$ and the limit energy:

$$\mathscr{W}(\boldsymbol{u}) = \int_{\mathcal{B}} \eta\left(\frac{\partial^2 u_2}{\partial x_1^2}\right)^2 + \chi\left(\frac{\partial \gamma}{\partial x_1}\right)^2 + \xi\,(\gamma - E_{12})^2 + \kappa\left(E_{12}\right)^2 \mathrm{d}V.$$

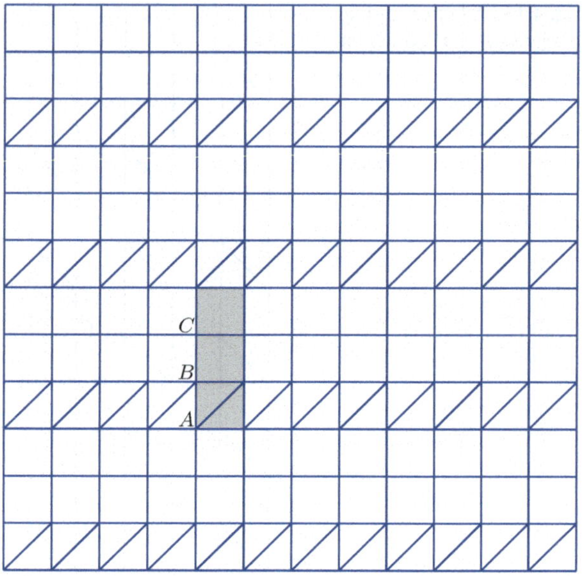

Fig. 4.14 A lattice leading to both strain-gradient and generalized continuum effects

Other Examples

The homogenization result established in this chapter is considering only structures lying in a 2D physical space with two vectors of periodicity in this space. It can be extended to the 3D case. It can also be extended to the case when a 3D (or 2D) structure has less than three (two) vectors of periodicity. The results enter then the framework of "dimension reduction." In those cases, we obtain effective plate, membrane or beam models. In this chapter, we have only presented 2D–2D examples. Different examples can be found in Abdoul-Anziz and Seppecher (2018b).

References

Abdoul-Anziz, H., & Seppecher, P. (2018a). Homogenization of periodic graph-based elastic structures. *Journal de l'Ecole polytechnique–Mathématiques*, *5*, 259–288.

Abdoul-Anziz, H., & Seppecher, P. (2018b). Strain gradient and generalized continua obtained by homogenizing frame lattices. *Mathematics and Mechanics of Complex Systems*, *6*(3), 213–250.

Abdoul-Anziz, H., Bellis, C., & Seppecher, P. (2019). Homogenization of frame lattices leading to second gradient models coupling classical strain and strain-gradient terms. *Mathematics and Mechanics of Solids*, *24*(12), 3976–3999.

Allaire, G. (1992). Homogenization and two-scale convergence. *SIAM Journal on Mathematical Analysis*, *23*(6), 1482–1518.

Braides, A. (2002). *Γ-convergence for Beginners*. Oxford: Oxford University Press.

Dal Maso, G. (1993). *An Introduction to Γ-Convergence*. Boston: Birkhäuser.

d'Alembert, J. L. R. (1743). *Traité de dynamique*.

dell'Isola, F., & Seppecher, P. (1997). Edge contact forces and quasi-balanced power. *Meccanica*, *32*(1), 33–52. ISSN 1572-9648. https://doi.org/10.1023/A:1004214032721.

dell'Isola, F., Seppecher, P., & Madeo, A. (2012). How contact interactions may depend on the shape of Cauchy cuts in nth gradient continua: approach "à la D'Alembert". *Zeitschrift für angewandte Mathematik und Physik*, *63*(6), 1119–1141. ISSN 1420-9039. https://doi.org/10.1007/s00033-012-0197-9.

Dell'Isola, F., Seppecher, P., & Della Corte, A. (2015). The postulations à la D'Alembert and à la Cauchy for higher gradient continuum theories are equivalent: a review of existing results. *Proceedings of the Royal Society A: Mathematical, Physical and Engineering Sciences*, *471*(2183), 20150415.

Germain, P. (1973a). *Cours de Mécanique des Milieux Continus*. Masson, Paris: Tome I: Théorie générale.

Germain, P. (1973b). La méthode des puissances virtuelles en mécanique des milieux continus. *Journal de Mécanique*, *12*, 236–274.

Milton, G. (1992). Composite materials with poisson's ratio close to -1. *Journal of the Mechanics and Physics of Solids*, *40*, 1105–1137. https://doi.org/10.1016/0022-5096(92)90063-8.

Milton, G. W., & Cherkaev, A. V. (1995). Which elasticity tensors are realizable? *Journal of Engineering Materials and Technology*, *117*(4), 483–493.

Nguetseng, G. (1989). A general convergence result for a functional related to the theory of homogenization. *SIAM Journal on Mathematical Analysis*, *20*(3), 608–623.

Noll, W. (1973). Lectures on the foundations of continuum mechanics and thermodynamics. *Archive for Rational Mechanics and Analysis*, *52*(1), 62–92.

Chapter 5
Strain Gradient Plasticity: Theory and Implementation

Lorenzo Bardella and Christian F. Niordson

Abstract This chapter focuses on the foundation and development of various *higher-order* strain gradient plasticity theories, and it also provides the basic elements for their finite element implementation. To this aim, we first refer to experiments exhibiting size-effects in metals and explain them by resorting to the concept of *geometrically necessary dislocations*. We then bring this concept to the continuum level by introducing Nye's dislocation density tensor and by postulating the existence of higher-order stresses associated with dislocation densities. This provides the motivation for the development of higher-order strain gradient plasticity theories. For this purpose, we adopt the generalized principle of virtual work, initially illustrated for conventional crystal plasticity and subsequently extended to both crystal and phenomenological strain gradient plasticity theories.

Keywords Micron-scale metal plasticity · Geometrically necessary dislocations · Dislocation density tensor · Strain gradient crystal plasticity · Strain gradient plasticity

L. Bardella (✉)
Department of Civil, Environmental, Architectural Engineering and Mathematics, University of Brescia, Brescia, Italy
e-mail: lorenzo.bardella@ing.unibs.it

C. F. Niordson
Department of Mechanical Engineering, Technical University of Denmark, Kgs., Lyngby, Denmark
e-mail: cn@mek.dtu.dk

A. Bertram and S. Forest (eds.), *Mechanics of Strain Gradient Materials*,
CISM International Centre for Mechanical Sciences 600,
https://doi.org/10.1007/978-3-030-43830-2_5

Introduction

First, this chapter goes through essential experiments, on metal samples of size range from a few tens of nanometers to a few tens of micrometers, that exhibit non-trivial size-effects in the form of an increase in yield point and hardening with diminishing size. In section "Experimental Motivation", this will provide the most important reason for the development of size-dependent plasticity theories. The kinematical foundation for metal plasticity is then laid out in section "Kinematics", with emphasis on Nye's dislocation density tensor and its relation to geometrically necessary dislocations (GNDs). As a fundamental tool for the development of strain gradient plasticity theories, section "Conventional Crystal Plasticity" introduces the generalized principle of virtual work (PVW). On this basis and with specific emphasis on the concept of GNDs, section "Micromechanical Motivations for the Introduction of Higher-Order Stresses" provides further reasons for the need of developing extensions of conventional plasticity to strain gradient theories involving higher-order stresses that are work conjugate to appropriate kinematic fields characterizing the spatial gradient of the plastic flow. Section "Strain Gradient Crystal Plasticity" presents in detail a strain gradient crystal plasticity (SGCP) theory, and finally, section "Phenomenological Strain Gradient Plasticity" lays out a foundation for phenomenological strain gradient plasticity (SGP) theories, also accounting for finite strains and focusing on details of the numerical implementation through the finite element (FE) method. In presenting each theory, we pay special attention to the possible constitutive assumptions for the higher-order stresses, that can be either recoverable or dissipative in nature, and both of these types of stress may be needed in a model in order to predict the size-effects experimentally observed.

Notation

We use index notation referred to an orthonormal system of coordinates, except when otherwise specified. For clarity, we drop indices and employ boldface letters for tensors in some cases when tensorial operations are not involved.

Experimental Motivation

Size-effects in metals have been well known since the discovery of the Hall–Petch effect (Hall 1951; Petch 1953), which describes how the yield stress of a metal increases with decreasing grain size. Beginning in the 1990s, similar micron-scale size-effects in metal plasticity have been confirmed in many experiments under simple overall deformation such as bending (Stölken and Evans 1998) and torsion (Fleck et al. 1994), as well as for complex deformation states such as indentation (Stelmashenko et al. 1993; Ma and Clarke 1995). Popularly, the observed effects can be summarized as *"smaller is stronger."*

The conventional theory for torsion of rods predicts that the torque, Q, should scale proportionally with the cube of the radius, a, and that the normalized torque, Q/a^3, is a unique function of the shear strain at the surface. Although this is accurate on larger scales, Fleck et al. (1994) showed that thin crystalline copper wires gain increased plastic resistance on the scale of about tens of microns and below. The more recent results of Liu et al. (2013) have confirmed these findings.

Similarly, conventional continuum theories suggest that the bending moment, M, for a beam made from a material possessing no size dependence, should scale proportionally with the width b and the square of the height h. Furthermore, the normalized bending moment $M/(bh^2)$ should be a unique function of the maximum longitudinal strain. Although this is accurate on larger scales, Stölken and Evans (1998) showed that a strong size dependence in the plastic deformation regime exists for thin nickel beams on the scale of about tens of microns and below. Confirmation of this behavior has been later provided, for instance, by Moreau et al. (2005).

Self-similar indenters, such as Berkovich and Vickers indenters, may also be used to quantify size-effects on the micron-scale. In a scale-independent material, hardness measurements with self-similar indenters should yield size-independent measurements, as there is only one independent length scale characterizing the deformation state, such as the indent depth or indenter diagonal. However, as shown by Stelmashenko et al. (1993) for tungsten and Ma and Clarke (1995) for silver, size-dependent hardness measurements are obtained for indenter diagonals of about 10 microns and below.

It is generally accepted that the size-effects confirmed in the above experiments are associated with large plastic strain gradients (Hutchinson 2000), which must be accommodated by a density of the so-called geometrically necessary dislocations (GNDs) (Ashby 1970). The Hall–Petch effect may also be interpreted in the context of large plastic strain gradients. Since grain boundaries act as obstacles to dislocation movement, dislocations pile up thus leaving a plastic strain gradient near grain boundaries. Hence, decreasing the distance between grain boundaries by diminishing grain size implies larger plastic strain gradients at a given overall plastic strain.

For tension of thin copper wires, where the overall strain gradient vanishes, Fleck et al. (1994) showed that, differently from torsion, the micron-scale size-effect was negligible. This further supports the connection between micron-scale size-effects in metals and plastic strain gradients.

Although pure tension does not lead to overall strain gradients, the blockage of dislocations at interfaces can induce large plastic strain gradients and associated size-effects even under homogeneous overall deformation. This has been confirmed in experiments by Kraft et al. (2000) where thin copper films deposited on an elastic substrate were found to exhibit a significant size-effect under overall homogeneous deformation due to dislocation blockage at the interface. Similarly, Xiang and Vlassak (2006) quantified a size-effect due to plastic strain gradients induced by applying an ultra-thin passivation layer blocking dislocations on one side of a thin copper film deformed homogeneously in a plane strain bulge test.

Since the mid-1990s, several size-scale-enriched continuum theories have emerged with the aim of accurately modeling the observed size-effects in metals. While some retain the lower-order structure of a conventional boundary value problem such as the framework proposed by Acharya and Bassani (2000), others employ higher-order frameworks, with higher-order work conjugate pairs providing enriched modeling capabilities for dislocation blockage at interfaces (Fleck et al. 1994, 2015; Fleck and Hutchinson 1997, 2001; Polizzotto et al. 1998; Gao et al. 1999; Gurtin 2002, 2004; Forest and Sievert 2003; Gudmundson 2004; Evers et al. 2004; Bardella 2006; Gurtin and Anand 2007, 2009; Kuroda and Tvergaard 2008; Polizzotto 2009; Ertürk et al. 2009).

In this chapter, we present a class of modern higher-order gradient plasticity theories. We begin by introducing the relevant kinematics involved in the small deformation theory of plasticity.

Kinematics

Basics of Linearized Kinematics

Within the small strains and rotations framework, the gradient of the displacement vector field admits the additive decomposition into its elastic, u^e_{ij}, and plastic, u^p_{ij}, parts

$$u_{i,j} = u^e_{ij} + u^p_{ij}$$

In crystal plasticity, the plastic distortion is defined as

$$u^p_{ij} = \sum_{\beta} \gamma^{(\beta)} s^{(\beta)}_i m^{(\beta)}_j \tag{5.1}$$

where the crystallographic unit vectors $s^{(\beta)}_i$ and $m^{(\beta)}_i$ denote, respectively, the slip direction and the slip plane normal of the slip system β where the plastic slip $\gamma^{(\beta)}$ occurs. As an example, face-centered cubic metals, such as copper, silver, and gold, have twelve slip systems, so that $\beta = 1, \ldots, 12$. Equation (5.1) implies that the trace of the plastic distortion vanishes, $u^p_{ii} = 0$, that is, no volumetric plastic strain occurs. We assume isochoric plastic deformation also in the case of phenomenological plasticity, which is a theory disregarding the details of individual plastic slip contributions to deformation, but rather accounts for the collected contributions of slip in a phenomenological manner.

The plastic distortion is conveniently split into its symmetric and skew-symmetric parts

$$\varepsilon^p_{ij} = \frac{1}{2}\left(u^p_{ij} + u^p_{ji}\right) \tag{5.2}$$

$$\vartheta_{ij}^{p} = \frac{1}{2}\left(u_{ij}^{p} - u_{ji}^{p}\right)$$

that are the plastic strain and the plastic spin, respectively. Hence, substitution of Eq. (5.1) into Eq. (5.2) leads to

$$\varepsilon_{ij}^{p} = \sum_{\beta} \gamma^{(\beta)} \mu_{ij}^{(\beta)}$$

where

$$\mu_{ij}^{(\beta)} = \frac{1}{2}\left(s_i^{(\beta)} m_j^{(\beta)} + s_j^{(\beta)} m_i^{(\beta)}\right)$$

is the symmetric Schmid tensor.

Nye's Dislocation Density Tensor

Nye's dislocation density tensor (Nye 1953; Kröner 1962; Fleck and Hutchinson 1997; Arsenlis and Parks 1999) is the curl of the plastic distortion, thus providing a measure of the incompatibility of the plastic deformation field by

$$\alpha_{ij} = e_{jkl} u_{il,k}^{p} \tag{5.3}$$

with $e_{ijk} = (i - j)(j - k)(k - i)/2$, which, in the literature, is referred to as the Ricci-Curbastro (or Levi-Civita) permutation tensor or simply the alternating tensor.

Definition (5.3) ensues from the closure failure of a general plastic distortion circuit, which, by referring to the terminology of dislocation mechanics (see, e.g., Hull and Bacon 2001), provides a continuum measure of a *net* Burgers vector g_i^{net} through

$$g_i^{\text{net}} = \oint_C u_{ij}^{p} \mathrm{d}x_j = \int_S e_{jkl} u_{il,k}^{p} n_j^{(S)} \mathrm{d}A \equiv \int_S \alpha_{ij} n_j^{(S)} \mathrm{d}A \tag{5.4}$$

Here, x_i is the coordinate along the circuit C enclosing a surface S with right-handed unit normal $n_i^{(S)}$ and the second equality is obtained by applying Stokes' theorem. Relations (5.4) provide the interpretation of α_{ij} as the ith component of the net Burgers vector related to dislocations of line vector j piercing the surface S. As shown in some detail below when referring to *crystal* plasticity, this implies that Nye's tensor components with equal indices can be related to densities of *pure* screw dislocations, whereas components having unequal indices can be related to densities of *pure* edge dislocations (see, e.g., Hull and Bacon 2001). Moreover, note that Nye's tensor has zero divergence, $\alpha_{ij,j} = 0$, and this property is referred to as Frank's rule or conservation of the Burgers vector (Hull and Bacon 2001).

The matrix representation of Nye's tensor (5.3) can be expressed as

$$\alpha = \begin{bmatrix} u^p_{13,2} - u^p_{12,3} & u^p_{11,3} - u^p_{13,1} & u^p_{12,1} - u^p_{11,2} \\ u^p_{23,2} - u^p_{22,3} & u^p_{21,3} - u^p_{23,1} & u^p_{22,1} - u^p_{21,2} \\ u^p_{33,2} - u^p_{32,3} & u^p_{31,3} - u^p_{33,1} & u^p_{32,1} - u^p_{31,2} \end{bmatrix}$$

This representation is strictly related to the adopted orthonormal reference system.

Nye's Dislocation Density Tensor in Crystal Plasticity

Now, we focus on crystal plasticity, where Nye's tensor can be related to the concept of GNDs densities (Ashby 1970).

By combining definitions (5.3) and (5.1), Nye's tensor can be rewritten as

$$\alpha_{ij} = e_{jkl} \sum_{\beta} \gamma^{(\beta)}_{,k} s^{(\beta)}_i m^{(\beta)}_l \tag{5.5}$$

It is now convenient to introduce the unit vector $l^{(\beta)}_i$ for the transverse direction on the slip system, which is related to the other crystallographic directions through the vector product

$$m^{(\beta)}_i = e_{ijk} l^{(\beta)}_j s^{(\beta)}_k$$

Hence, we can rewrite Eq. (5.5) as

$$\begin{aligned} \alpha_{ij} = e_{ljk} e_{lpq} \sum_{\beta} \gamma^{(\beta)}_{,k} s^{(\beta)}_i l^{(\beta)}_p s^{(\beta)}_q &= (\delta_{jp}\delta_{kq} - \delta_{jq}\delta_{kp}) \sum_{\beta} \gamma^{(\beta)}_{,k} s^{(\beta)}_i l^{(\beta)}_p s^{(\beta)}_q \\ &= \sum_{\beta} \gamma^{(\beta)}_{,k} s^{(\beta)}_i \left(s^{(\beta)}_k l^{(\beta)}_j - s^{(\beta)}_j l^{(\beta)}_k \right) \end{aligned} \tag{5.6}$$

where the second equality of Eq. (5.6) makes use of the $e - \delta$ identity (see, e.g., Malvern 1969, p. 23).

On the one hand, we may define the crystallographic third-order tensor $N^{(\beta)}_{ikj} = s^{(\beta)}_i (s^{(\beta)}_k l^{(\beta)}_j - s^{(\beta)}_j l^{(\beta)}_k)$ such that (Bardella 2006)

$$\alpha_{ij} = \sum_{\beta} \gamma^{(\beta)}_{,k} N^{(\beta)}_{ikj} \tag{5.7}$$

On the other hand, the introduction of the densities of *pure* edge dislocations,

$$\rho^{(\beta)}_{\perp} = \gamma^{(\beta)}_{,i} s^{(\beta)}_i , \tag{5.8}$$

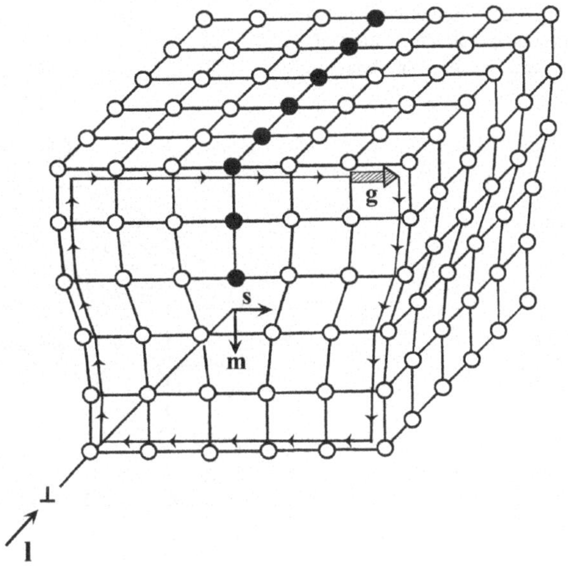

Fig. 5.1 Schematic for the atomistic characterization of an edge dislocation, consisting of a half plane of atoms (black circles) in an otherwise regular crystal lattice. Adapted from Bardella (2020)

and *pure* screw dislocations,

$$\rho_{\odot}^{(\beta)} = -\gamma_{,i}^{(\beta)} l_i^{(\beta)} , \tag{5.9}$$

allows us to rewrite Nye's dislocation density tensor (Arsenlis and Parks 1999) as

$$\alpha_{ij} = \sum_{\beta} s_i^{(\beta)} \left(\rho_{\perp}^{(\beta)} l_j^{(\beta)} + \rho_{\odot}^{(\beta)} s_j^{(\beta)} \right) \tag{5.10}$$

The well-posedness of definitions (5.8)–(5.10) can be understood from Figs. 5.1 and 5.2, where, for a cubic crystal, we display the Burgers circuits whose closure failure defines the Burgers vector g_i characterizing *pure* edge and screw dislocations (Burgers 1939), along with the crystallographic directions s_i, m_i, and l_i. The magnitude of the Burgers vector corresponds to an interatomic spacing. For edge dislocations (Fig. 5.1), the slip direction s_i both is normal to the dislocation line and coincides with the direction of the plastic slip gradient (as expressed by Eq. (5.8)). For screw dislocations (Fig. 5.2), the slip direction s_i coincides with the dislocation line, whereas the direction of the plastic slip gradient is the transverse direction l_i (as expressed by Eq. (5.9)).

Hence, given the property of Nye's tensor deduced from relations (5.4), inspection of definitions (5.8)–(5.10) in comparison with Figs. 5.1 and 5.2 allows one to deduce that $\rho_{\perp}^{(\beta)}$ and $\rho_{\odot}^{(\beta)}$ represent the densities of *pure* edge and *pure* screw dislocations, respectively, for the slip system β (Arsenlis and Parks 1999). Here, the adjective "pure" refers to the ideal characterization illustrated in Figs. 5.1 and 5.2, whereas

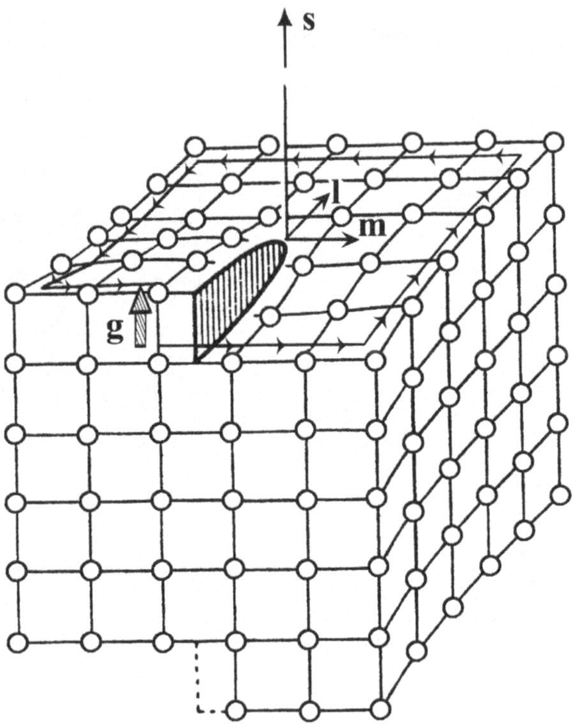

Fig. 5.2 Schematic for the atomistic characterization of a screw dislocation. Adapted from Bardella (2020)

real dislocations consist of loops, in general having, pointwise, both edge and screw components (Hull and Bacon 2001).

Moreover, relations (5.4) imply that Nye's tensor accounts only for densities of *excess* dislocations, also referred to as misfit dislocations or GNDs, in contrast to statistically stored dislocations (SSDs). As demonstrated by Ashby (1970) and deducible from the experiments illustrated in section "Experimental Motivation", GNDs are responsible for the size-dependent behavior here of interest. The key issue in strain gradient plasticity is the connections between GNDs and the gradient of the plastic distortion.

As further information related to Figs. 5.1 and 5.2, we add that dislocations mostly move by glide, which is a dissipative process and occurs on a plane containing both the dislocation line and the Burgers vector.

Finally, from Eqs. (5.6) and (5.9) one deduces that the trace of Nye's tensor provides the sum of the densities of pure screw dislocations as follows

$$\alpha_{ii} = -\sum_{\beta} \gamma_{,i}^{(\beta)} t_i^{(\beta)} = \sum_{\beta} \rho_{\odot}^{(\beta)}$$

We conclude that gradients of slip are associated with GNDs that may be characterized by Nye's tensor. In view of experiments exhibiting size-effects whenever large plastic strain gradients evolve, it points to GNDs being central for quantifying size-effects in metals, as will be further discussed in section "Micromechanical Motivations for the Introduction of Higher-Order Stresses".

Now, within the framework of phenomenological plasticity, we focus on the torsion problem to provide a meaningful example of characterization of the plastic distortion incompatibility through Nye's tensor. The following results will be employed in section "Distortion Gradient Plasticity" in order to illustrate some of the most relevant modeling capabilities of higher-order strain gradient plasticity.

The Kinematics of the Torsion of Thin Wires

Within the phenomenological theory of plasticity, the problem of the torsion of thin wires is conveniently studied in a cylindrical reference system. In particular, by denoting with z the axis of an isotropic wire of circular cross section, and with r and θ the radial and azimuthal coordinates, respectively, the problem symmetries allow us to write the displacement in terms of the applied twist per unit length, κ, as

$$u_\theta = \kappa z r\ , \qquad u_r = u_z = 0 \tag{5.11}$$

The sole nonvanishing components of the plastic distortion are

$$u^p_{\theta z}(r) = \varepsilon^p(r) + \vartheta^p(r) \quad \text{and} \quad u^p_{z\theta}(r) = \varepsilon^p(r) - \vartheta^p(r)$$

where we have simplified the notation by dropping indices as follows

$$\varepsilon^p(r) \equiv \varepsilon^p_{\theta z}(r) \quad \text{and} \quad \vartheta^p(r) \equiv \vartheta^p_{\theta z}(r) = -\vartheta^p_{z\theta}(r) \tag{5.12}$$

Now, Nye's tensor may be computed by appropriately extending the curl of a vector field in cylindrical coordinates (see, e.g., Malvern 1969). By denoting with $e_i^{(r)}$, $e_i^{(\theta)}$, and $e_i^{(z)}$ the unit vectors along the coordinate axes r, θ, and z, respectively, we obtain

$$\begin{aligned}\alpha_{ij} = \frac{\partial[e_i^{(\theta)}(\varepsilon^p + \vartheta^p)]}{r\partial\theta} e_j^{(r)} - \frac{\partial[e_i^{(\theta)}(\varepsilon^p + \vartheta^p)]}{\partial r} e_j^{(\theta)} \\ - \frac{\partial[e_i^{(z)}(\varepsilon^p - \vartheta^p)]}{\partial z} e_j^{(r)} + \frac{\partial[e_i^{(z)}(\varepsilon^p - \vartheta^p) r]}{r\partial r} e_j^{(z)}\end{aligned}$$

By accounting for the relation $\partial e_i^{(\theta)}/\partial\theta = -e_i^{(r)}$ and by employing the prime notation for the radial derivative, we finally have (Bardella and Panteghini 2015)

$$\alpha = \begin{bmatrix} -\dfrac{\varepsilon^p + \vartheta^p}{r} & 0 & 0 \\ 0 & -(\varepsilon^p)' - (\vartheta^p)' & 0 \\ 0 & 0 & \dfrac{\varepsilon^p - \vartheta^p}{r} + (\varepsilon^p)' - (\vartheta^p)' \end{bmatrix} \tag{5.13}$$

This representation of Nye's tensor indicates that in this torsion problem only densities of pure screw GNDs are relevant. It is interesting to observe that the first invariant of Nye's tensor (5.13) depends on the plastic spin only according to

$$\alpha_{ii} = -2\left[\frac{\vartheta^p}{r} + (\vartheta^p)'\right] \tag{5.14}$$

Moreover, for later use in setting the constitutive laws, it is relevant to consider the deviatoric part of Nye's tensor. Here and henceforth, we indicate deviatoric second-order tensors by adding the superscript d to the whole tensor, such that

$$\alpha^{\mathrm{d}}_{ij} = \alpha_{ij} - \frac{\alpha_{kk}}{3}\delta_{ij}$$

where δ_{ij} is the second-order unit tensor (Kronecker delta). Then, α^{d}_{ij} may be expressed by

$$\begin{bmatrix} -\dfrac{\varepsilon^p}{r} - \dfrac{\vartheta^p}{3r} + \dfrac{2}{3}(\vartheta^p)' & 0 & 0 \\ 0 & \dfrac{2\vartheta^p}{3r} - (\varepsilon^p)' - \dfrac{1}{3}(\vartheta^p)' & 0 \\ 0 & 0 & \dfrac{\varepsilon^p}{r} - \dfrac{\vartheta^p}{3r} + (\varepsilon^p)' - \dfrac{1}{3}(\vartheta^p)' \end{bmatrix}$$

such that its second invariant may be written as

$$|\alpha^{\mathrm{d}}| \equiv \sqrt{\alpha^{\mathrm{d}}_{ij}\alpha^{\mathrm{d}}_{ij}} = \left\{ 2\left(\frac{\varepsilon^p}{r}\right)^2 + \frac{2}{3}\left(\frac{\vartheta^p}{r}\right)^2 + 2\frac{\varepsilon^p}{r}[(\varepsilon^p)' - (\vartheta^p)'] \right.$$
$$\left. -2\frac{\vartheta^p}{r}\left[(\varepsilon^p)' + \frac{1}{3}(\vartheta^p)'\right] + 2[(\varepsilon^p)']^2 + \frac{2}{3}[(\vartheta^p)']^2 \right\}^{1/2} \tag{5.15}$$

Conventional Crystal Plasticity

Principle of Virtual Work

We consider a body occupying a space region Ω, whose external surface $S = S_T \cup S_U$ has outward unit normal n_i. The body is subjected to the following known fields: the body force b_i in Ω, the traction vector T^0_i on S_T, and the displacement u^0_i on S_U.

In any subregion $\Pi \subset \Omega$, the standard forms of the internal virtual work, $\mathcal{W}_i$, and external virtual work, $\mathcal{W}_e$, respectively, read

$$\mathcal{W}_i = \int_\Pi \sigma_{ij}\delta\varepsilon_{ij}\mathrm{d}V \tag{5.16}$$

$$\mathcal{W}_e = \int_\Pi b_i\delta u_i\mathrm{d}V + \int_{\partial\Pi} T_i\delta u_i\mathrm{d}A \tag{5.17}$$

where T_i is the traction vector acting on the internal surface $\partial\Pi$ and, here and henceforth, $\delta\epsilon$ denotes any *kinematically admissible variation* of the field ϵ (Lanczos 1970).

Toward the development of a *generalized* principle of virtual work (PVW) useful to obtain the balance equations governing strain gradient plasticity theories, we non-conventionally introduce, for each slip system, the stress $q^{(\beta)}$ and modify the internal virtual work (5.16) as follows, in which δu_i and $\delta\gamma^{(\beta)}$, $\forall\beta$, are assumed to be *independent* variations:

$$\begin{aligned}\mathcal{W}_i &= \int_\Pi \Big(\underbrace{\sigma_{ij}\delta\varepsilon^e_{ij}}_{\text{recoverable}} \underbrace{+\sum_\beta q^{(\beta)}\delta\gamma^{(\beta)}}_{\text{dissipative}} \Big)\mathrm{d}V \\ &= \int_\Pi \Big(\sigma_{ij}\left(\delta\varepsilon^e_{ij} + \delta\varepsilon^p_{ij}\right) + \sum_\beta (q^{(\beta)} - \tau^{(\beta)})\delta\gamma^{(\beta)}\Big)\mathrm{d}V \\ &= \int_\Pi \Big(\sigma_{ij}\delta\varepsilon_{ij} + \sum_\beta (q^{(\beta)} - \tau^{(\beta)})\delta\gamma^{(\beta)}\Big)\mathrm{d}V \end{aligned} \tag{5.18}$$

where

$$\tau^{(\beta)} = \sigma_{ij}s_i^{(\beta)}m_j^{(\beta)} = \sigma_{ij}\mu_{ij}^{(\beta)} \;\; \forall\beta \tag{5.19}$$

is the resolved shear stress (also referred to as the Schmid stress).

Now, by postulating the PVW, that is

$$\mathcal{W}_i = \mathcal{W}_e\,, \tag{5.20}$$

with the definitions (5.18) and (5.17), by integrating by parts, and by applying the divergence theorem, we obtain

$$\int_\Pi \Big[(-\sigma_{ij,j} - b_i)\delta u_i + \sum_\beta (q^{(\beta)} - \tau^{(\beta)})\delta\gamma^{(\beta)}\Big]\mathrm{d}V = \int_{\partial\Pi} (T_i - \sigma_{ij}n_j)\delta u_i\mathrm{d}A \tag{5.21}$$

which must hold $\forall\delta u_i, \delta\gamma^{(\beta)}$. Hence, standard arguments of calculus of variation (Lanczos 1970) lead to the balance equations

$$\sigma_{ij,j} + b_i = 0 \quad \text{in} \quad \Pi \tag{5.22}$$

$$\sigma_{ij} n_j = T_i \quad \text{on} \quad \partial\Pi \tag{5.23}$$

along with the *Schmid law*, that has to hold when plasticity occurs

$$q^{(\beta)} - \tau^{(\beta)} = 0 \quad \forall \beta \quad \text{in} \quad \Omega \tag{5.24}$$

Here, the Schmid law is obtained as a *pseudo*-balance equation (Del Piero 2009), whereas it is classically introduced at the constitutive level as the yield criterion together with the law of variation of $q^{(\beta)}$ with respect to the plastic flow internal variable(s). We anticipate that in higher-order strain gradient plasticity pseudo-balance equations are differential equations, although this is not the case in conventional plasticity.

From (5.23), the conventional static boundary condition reads

$$\sigma_{ij} n_j = T_i^0 \quad \text{on} \quad S_T \tag{5.25}$$

while u_i is known on S_U where its variation δu_i vanishes.

In the following, for the sake of simplicity, we neglect body forces, such that the conventional balance Eq. (5.22) becomes

$$\sigma_{ij,j} = 0 \quad \text{in} \quad \Omega \tag{5.26}$$

Constitutive Relations

In order to complete the conventional crystal plasticity theory, constitutive equations must be established. This is usually done on the basis of simple experiments providing the uniaxial response of the material augmented by assumptions on the energy expenditure in the material. In crystal plasticity, one may assume that the incremental internal energy density can be split into an elastic part given in terms of the elastic strain increment that works against the Cauchy stress, and a part associated with the various plastic slip quantities $\dot{\gamma}^{(\beta)}$ working against their conjugate stress quantities $q^{(\beta)}$ according to

$$\dot{\mathcal{U}} = \sigma_{ij} \dot{\varepsilon}_{ij}^e + \sum_\beta q^{(\beta)} \dot{\gamma}^{(\beta)}$$

Here and henceforth, $\dot{\epsilon} \equiv \mathrm{d}\epsilon / \mathrm{d}t$ is the rate of the field ϵ. One may further specify the energy that is stored in terms of the Helmholtz free energy density, here assuming no contribution from plastic slips:

$$\mathcal{E} = \frac{1}{2} \varepsilon_{ij}^e L_{ijkl} \varepsilon_{kl}^e \tag{5.27}$$

where L_{ijkl} is the fourth-order elasticity tensor. The Cauchy stress, which is work conjugate to the elastic strain, may then be obtained from

$$\sigma_{ij} = \frac{\mathrm{d}\mathcal{E}}{\mathrm{d}\varepsilon_{ij}^{e}} = L_{ijkl}\varepsilon_{kl}^{e} \tag{5.28}$$

and it follows that the energy dissipated is given by

$$\dot{\mathcal{U}} - \dot{\mathcal{E}} = \sum_{\beta} q^{(\beta)}\dot{\gamma}^{(\beta)} > 0 \quad \forall \dot{\underline{\gamma}} \neq \underline{0}$$

where $\underline{\gamma}$ is the vector containing all the plastic slips and the inequality ensures fulfillment of the second law of thermodynamics. This fundamental law can be satisfied by simply imposing

$$\mathrm{sgn}\,(q^{(\beta)}) = \mathrm{sgn}\,(\dot{\gamma}^{(\beta)}) \quad \forall \beta$$

Rate-Independent Conventional Crystal Plasticity

In the rate-independent case, the constitutive laws are usually completed (among the earliest pivotal contributions, see Bishop and Hill 1951) by the introduction of the following yield function

$$\varphi^{(\beta)} = |\tau^{(\beta)}| - |q^{(\beta)}| \quad \forall \beta\,,$$

normality rule

$$\mathrm{sgn}\,(\tau^{(\beta)}) = \mathrm{sgn}\,(\dot{\gamma}^{(\beta)})\,,$$

Karush–Kuhn–Tucker (KKT) conditions

$$\varphi^{(\beta)} \leq 0\,, \quad \varphi^{(\beta)}\dot{\gamma}^{(\beta)} = 0 \quad \forall \beta\,, \tag{5.29}$$

consistency condition

$$\dot{\varphi}^{(\beta)}\dot{\gamma}^{(\beta)} = 0 \;\; \forall \beta \tag{5.30}$$

Finally, it must be defined how the "slip resistance" evolves with plastic flow, according to

$$|q^{(\beta)}| = |q^{(\beta)}(\hat{\underline{\gamma}})|$$

which may include both self and latent isotropic hardening through the "accumulated slip vector" $\hat{\underline{\gamma}} = \int |\dot{\underline{\gamma}}|\,\mathrm{d}t$ (see, e.g., Peirce et al. 1983). Here, we denote the yield stress, that is, the value of $|q^{(\beta)}|$ at incipient plasticity, with the symbol $\tau_0^{(\beta)} = |q^{(\beta)}(\underline{0})|$.

The above-defined yield criterion may lead to non-uniqueness in the activation of slip systems, which can be overcome by the development of regularized rate-independent criteria, such as that proposed by Gambin (1992). Another way to regularize the problem is to adopt a viscoplasticity framework, as introduced in the following, which also enables to address other special issues of higher-order strain gradient rate-independent plasticity.

Conventional Crystal Viscoplasticity

In the case of viscoplasticity, it is assumed that plasticity develops at any stress level. The plastic strain rate is given implicitly by an expression for the level of stress $q^{(\beta)}$ given in terms of the slips and the slip rates

$$q^{(\beta)} = q^{(\beta)}\left(\underline{\hat{\gamma}}, \underline{\dot{\gamma}}\right)$$

to be substituted in the Schmid law (5.24).

In most cases, the dependence on the plastic slips (and the deformation history) is such that

$$q^{(\beta)} = g^{(\beta)}(\underline{\hat{\gamma}}) V(|\dot{\gamma}^{(\beta)}|) \,\mathrm{sgn}\,(\dot{\gamma}^{(\beta)}) \quad \forall \beta$$

in which $g^{(\beta)}(\underline{\hat{\gamma}})$ governs self and latent isotropic hardening, whereas the viscoplastic potential $V(|\dot{\gamma}^{(\beta)}|)$ accounts for the rate dependence.

A widely used power law form is given by (Needleman 1988)

$$V(|\dot{\gamma}^{(\beta)}|) = \left(\frac{|\dot{\gamma}^{(\beta)}|}{\dot{\gamma}_0^{(\beta)}}\right)^m$$

with $\dot{\gamma}_0$ and m denoting positive material parameters referred to as the reference slip rate and the rate sensitivity exponent, respectively. The limit $m \to 0$ leads to rate independence.

Alternatively, a regularization of the unit step function may be used as expressed by (Bardella and Panteghini 2015)

$$V(|\dot{\gamma}^{(\beta)}|) = \begin{cases} \dfrac{|\dot{\gamma}^{(\beta)}|}{2\dot{\gamma}_0} & \text{if } \dfrac{|\dot{\gamma}^{(\beta)}|}{\dot{\gamma}_0} \leq 1 \\ \dfrac{1}{2(P-1)}\left[P - 2 + \left(\dfrac{|\dot{\gamma}^{(\beta)}|}{\dot{\gamma}_0}\right)^{P-1}\right] & \text{if } \dfrac{|\dot{\gamma}^{(\beta)}|}{\dot{\gamma}_0} > 1 \end{cases} \tag{5.31}$$

with $\dot{\gamma}_0 > 0$ and $P \geq 0$ denoting two material parameters, the latter assuming the role of *rate sensitivity exponent*. The choice $P = 0$ has been successfully adopted in Panteghini and Bardella (2016, 2018), Martínez-Pañeda et al. (2019), and Panteghini et al. (2019) to efficiently approximate rate independence by additionally setting a suitably small value of $\dot{\gamma}_0$.

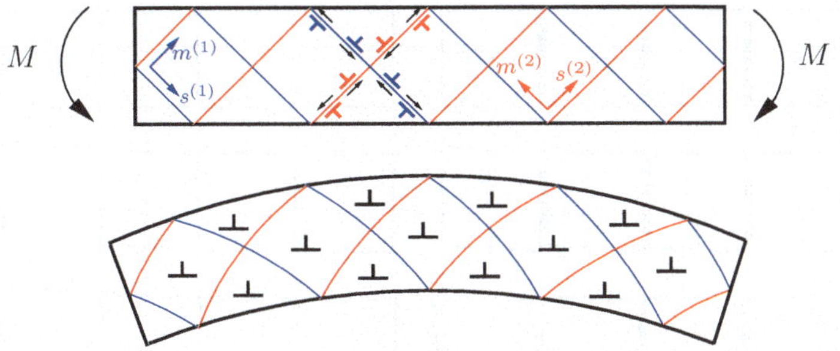

Fig. 5.3 Purely plastic deformation of a bent beam endowed with two symmetrically oriented slip systems

Micromechanical Motivations for the Introduction of Higher-Order Stresses

As an illustration of the link between plastic strain gradients and GNDs, consider Fig. 5.3. Here, a beam with two symmetrically oriented slip systems is loaded by a bending moment. By neglecting elasticity, the plastically deformed beam must have undergone a net flux of material from the lower (shorter) surface to the upper (longer surface), due to plastic incompatibility. This can be realized by a slip gradient along the slip planes, and it results in a net Burgers vector consistent with a uniform edge dislocation density which is proportional to the gradient of plastic strain through the beam thickness.

A further example is provided in Fig. 5.4, which shows a rectangular domain split in five sub-domains each envisioned to undergo uniform slip. The first row illustrates the undeformed geometry. In the second row, we show the geometry for a net slip gradient opposite the direction of the slip plane normal, $m^{(1)}$, whereas in the third row the net gradient of slip is in the slip direction, $s^{(2)}$. The right column displays the entire structure, whereas the left column shows the structure subdivided into the five domains that undergo uniform deformation. Imagine sources at the central line of the structure emitting zero units of slip in the left sub-domain, one unit of slip in the second domain, two units in the third domain, and henceforth. For the case with a net slip gradient in the direction of the slip normal (second row), the dislocations may exit the sub-domains, which fit compatibly together as shown to the right. However, for the case with a net slip gradient in the direction of the slip direction, a net density of dislocation must be stored, leading to an incompatible plastic deformation field that must be accommodated by elastic deformations.

By focusing on dislocations *at rest*, the distortions represented in Figs. 5.1 and 5.2 are associated with internal stress fields that become important when the dislocation density is high. The *long-range* stress field associated with SSDs densities vanishes because of the statistical nature of SSDs, leading to contributions that cancel out far from these dislocation densities, at a relatively small distance within the material.

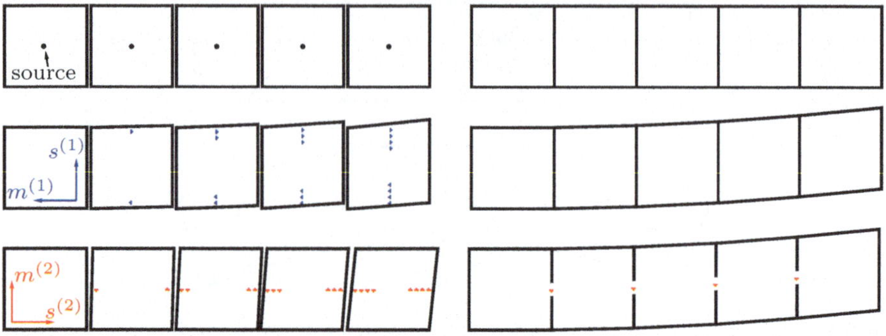

Fig. 5.4 Two ways of accommodating non-uniform slip in a rectangular domain

Instead, the stress fields due to GNDs sum up, because these *excess* dislocation densities contribute with the same sign, as illustrated in the foregoing examples concerned with Figs. 5.3 and 5.4. Hence, GND densities give rise to long-range stresses that are relevant at the micron-scale.

In higher-order SGP, a common way to incorporate the recoverable (energetic) contribution of GNDs (i.e., their contribution *at rest*) consists of introducing an addition to the Helmholtz free energy density referred to as the *defect energy* $\mathcal{D}$ (Gurtin 2002) that should be a function of the *current* value of the GND densities, to be represented by a suitable continuum measure of the plastic distortion incompatibility. Hence, in this context, the free energy density (5.27) becomes

$$\mathcal{E} = \frac{1}{2}\varepsilon^e_{ij} L_{ijkl}\varepsilon^e_{kl} + \mathcal{D} \tag{5.32}$$

Among the various options investigated in the literature, here we mention four of them. The defect energy may be assumed to be a function of Nye's dislocation density tensor, $\mathcal{D} = \mathcal{D}(\boldsymbol{\alpha})$, as formerly proposed by Gurtin (2002, 2004) (and adopted in the following sections "Strain Gradient Crystal Plasticity" and "Distortion Gradient Plasticity"), or, in crystal plasticity, even to the entire set of dislocation densities, $\mathcal{D} = \mathcal{D}(\rho^{(1)}_{\perp}, \rho^{(1)}_{\perp}, \ldots, \rho^{(1)}_{\odot}, \rho^{(2)}_{\odot}, \ldots)$, as for instance initially put forward by Nicola et al. (2005). Otherwise, aiming at a simpler modeling in phenomenological plasticity, the primal higher-order variable entering $\mathcal{D}$ may be chosen to be either the plastic strain gradient, $\mathcal{D} = \mathcal{D}(\nabla\boldsymbol{\varepsilon}^p)$, as in Gudmundson (2004) and Fleck and Willis (2009b), or even the gradient of the second invariant of the plastic strain, $\mathcal{D} = \mathcal{D}(\nabla\varepsilon^p_{\text{eq}})$ (with $\varepsilon^p_{\text{eq}} = \sqrt{(2/3)\varepsilon^p_{ij}\varepsilon^p_{ij}}$), as chosen in Fleck and Willis (2009a) to follow Fleck and Hutchinson (2001) in pursuing the most straightforward extension to SGP of conventional von Mises plasticity theory.

The gradient of the defect energy with respect to its primal higher-order variable provides the so-called *defect stress*. Hence, by referring to the foregoing examples, the defect stress may be a second-order tensor, a set of scalars, a third-order tensor, or a vector, depending on whether it is conjugate to $\boldsymbol{\alpha}$, $\{\rho_{\perp}^{(1)}, \rho_{\perp}^{(1)}, \ldots, \rho_{\odot}^{(1)}, \rho_{\odot}^{(2)}, \ldots\}$, $\nabla\boldsymbol{\varepsilon}^p$, or $\nabla\varepsilon_{\text{eq}}^p$, respectively. We remark that, in SGP, higher-order stresses do not necessarily have a higher tensorial order than that of the Cauchy stress.

It is well recognized that a quadratic defect energy is useful to model a size-effect in the form of an increase of strain hardening with diminishing size. However, *less-than-quadratic* defect energies may be tuned to provide a conspicuous increase of defect stress for a relatively small increment of the primal variable when this variable is small. In such a case, the predicted size-effect consists of an energetic strengthening, that is, an apparent increase of the yield point with diminishing size. Important classes of less-than-quadratic defect energy are, for instance, the "extreme" one-homogeneous energy (Conti and Ortiz 2005; Ohno and Okumura 2007; Garroni et al. 2010) and the logarithmic energy (Svendsen and Bargmann 2010; Forest and Guéninchault 2013; Bardella and Panteghini 2015; Wulfinghoff et al. 2015).

The dissipation due to GNDs motion (along with their nucleation, multiplication, and annihilation) is included in the continuum modeling in a more phenomenological way, as illustrated in the following.

Strain Gradient Crystal Plasticity

We present the SGCP theory developed by Gurtin (2002), Bardella (2006, 2007), Gurtin and Anand (2007), Gurtin et al. (2007), Bardella and Giacomini (2008). This theory assumes that GNDs enter the internal energy by contributing to both recoverable energy and dissipation. The higher-order recoverable contribution consists of a defect energy which is a function of Nye's tensor, $\mathcal{D}(\boldsymbol{\alpha})$. The higher-order dissipation is assumed to depend on the rates of the pure edge and screw dislocation densities along with the rate of the plastic slip projection onto the slip plane normal, denoted as

$$\rho_{\mathrm{M}}^{(\beta)} = \gamma_{,i}^{(\beta)} m_i^{(\beta)}$$

by extending the notation in Eqs. (5.8) and (5.9).

Hence, the internal virtual work on any subregion $\Pi \subset \Omega$ is defined as

$$\mathcal{W}_i = \int_\Pi \Bigg[\underbrace{\sigma_{ij}(\delta\varepsilon_{ij} - \delta\varepsilon_{ij}^p) + \zeta_{ij}\delta\alpha_{ij}}_{\text{energetic}} \\ \underbrace{+ \sum_\beta \left(q^{(\beta)}\delta\gamma^{(\beta)} + \omega_{\perp}^{(\beta)}\delta\rho_{\perp}^{(\beta)} + \omega_{\odot}^{(\beta)}\delta\rho_{\odot}^{(\beta)} + \omega_{\mathrm{M}}^{(\beta)}\delta\rho_{\mathrm{M}}^{(\beta)} \right)}_{\text{dissipative}} \Bigg] \mathrm{d}V \tag{5.33}$$

where ζ_{ij} is the defect stress obtained from the free energy density (5.32) as

$$\zeta_{ij} = \frac{\partial \mathcal{E}(\varepsilon^e, \alpha)}{\partial \alpha_{ij}} \equiv \frac{\mathrm{d}\mathcal{D}(\alpha)}{\mathrm{d}\alpha_{ij}} \tag{5.34}$$

and, for each slip system β, we have introduced the dissipative higher-order stresses $\omega_{\perp}^{(\beta)}, \omega_{\odot}^{(\beta)}$, and $\omega_{\mathrm{M}}^{(\beta)}$ that are work conjugate to $\dot{\rho}_{\perp}^{(\beta)}, \dot{\rho}_{\odot}^{(\beta)}$, and $\dot{\rho}_{\mathrm{M}}^{(\beta)}$, respectively. We note that the specific constitutive choices prescribing whether the higher-order stresses are energetic or dissipative are irrelevant in order to obtain the balance equations through the generalized PVW.

By substituting the kinematic relations introduced in section "Basics of Linearized Kinematics" in the internal virtual work (5.33) and by employing definition (5.19) for the resolved shear stress, integration by parts provides

$$\begin{aligned} \mathcal{W}_i = & \int_{\Pi} \Bigg[\sigma_{ij} \Big(\delta u_{i,j} - \sum_{\beta} \delta\gamma^{(\beta)} \mu_{ij}^{(\beta)} \Big) + \zeta_{ij} \sum_{\beta} \delta\gamma_{,k}^{(\beta)} N_{ikj}^{(\beta)} \\ & + \sum_{\beta} \Big(q^{(\beta)} \delta\gamma^{(\beta)} + \omega_{\perp}^{(\beta)} \delta\gamma_{,k}^{(\beta)} s_k^{(\beta)} - \omega_{\odot}^{(\beta)} \delta\gamma_{,k}^{(\beta)} l_k^{(\beta)} + \omega_{\mathrm{M}}^{(\beta)} \delta\gamma_{,k}^{(\beta)} m_k^{(\beta)} \Big) \Bigg] \mathrm{d}V \\ = & \int_{\Pi} \Big[-\sigma_{ij,j} \delta u_i - \sum_{\beta} \Big(\tau^{(\beta)} - q^{(\beta)} + \xi_{i,i}^{(\beta)} \Big) \delta\gamma^{(\beta)} \Big] \mathrm{d}V \\ & + \int_{\partial\Pi} \Big[\sigma_{ij} \delta u_i n_j + \sum_{\beta} \xi_i^{(\beta)} \delta\gamma^{(\beta)} n_i \Big] \mathrm{d}A \end{aligned} \tag{5.35}$$

where we have introduced the *microstress* $\xi_i^{(\beta)}$, accounting for all higher-order contributions

$$\xi_k^{(\beta)} = \zeta_{ij} N_{ikj}^{(\beta)} + \omega_{\perp}^{(\beta)} s_k^{(\beta)} - \omega_{\odot}^{(\beta)} l_k^{(\beta)} + \omega_{\mathrm{M}}^{(\beta)} m_k^{(\beta)} \tag{5.36}$$

On the basis of the boundary contribution in expression (5.35), one may define the external virtual work (in the absence of body forces) as

$$\mathcal{W}_e = \int_{\partial\Pi} \Big[T_i \delta u_i + \sum_{\beta} t^{(\beta)} \delta\gamma^{(\beta)} \Big] \mathrm{d}A$$

in which $t^{(\beta)}$ are non-conventional higher-order tractions. We note that this holds if any of the higher-order stresses $\omega_I^{(\beta)}$ ($I = \perp, \odot, \mathrm{M}$) is constitutively nonvanishing; otherwise, the non-conventional traction can be different, as discussed in section "On the Higher-Order Boundary Conditions".

The *generalized* PVW (5.20) then holds for all the admissible variations of displacement and plastic slip fields and delivers, along with the conventional balance Eqs. (5.26) and (5.25), the higher-order balance equation

$$q^{(\beta)} = \tau^{(\beta)} + \xi_{i,i}^{(\beta)} \quad \forall \beta \quad \text{in } \Omega \tag{5.37}$$

which is referred to as the *modified Schmid law* and has a contribution of the defect stress through the microstress (see Eq. (5.36)). Since the defect stress depends on Nye's tensor (5.7), the modified Schmid laws related to different slip systems are in general coupled, whereby Nye's tensor provides a natural source of higher-order latent hardening.

In this work, we always consider *homogeneous higher-order boundary conditions*.

For what concerns statics, such (natural) boundary conditions are referred to as *microfree* whereby modeling dislocations free to exit the body at plastically unconstrained external boundaries S_T^{dis}, where $t^{(\beta)} = 0\ \forall\beta$. Hence, the generalized PVW provides the microfree boundary conditions

$$\xi_i^{(\beta)} n_i = 0\ \forall\beta \quad \text{on} \quad S_T^{\mathrm{dis}} \tag{5.38}$$

Dual to the boundary condition (5.38), the non-conventional essential boundary conditions are referred to as *microhard* boundary conditions. They kinematically impose dislocation blockage (eventually leading to pileup) at plastically constrained external boundaries $S_U^{\mathrm{dis}} = S \setminus S_T^{\mathrm{dis}}$ through

$$\dot{\gamma}^{(\beta)} = 0\ \forall\beta \quad \text{on} \quad S_U^{\mathrm{dis}} \tag{5.39}$$

The microhard boundary conditions are of fundamental relevance as they may trigger inhomogeneous plastic flow in boundary value problems that would otherwise lead to spatially homogeneous plastic flow, such as in pure shear deformation and in plastically deforming thin films on elastic substrates. Note that we have written the microhard boundary conditions in terms of rate since loading histories including a switch from microfree to microhard boundary conditions (modeling the sudden formation of a passivation layer) are relevant in micron-scale plasticity (Fleck et al. 2014; Willis 2019).

Constitutive Relations

The Dissipative Behavior and the Dissipative Length Scales

The dissipative stresses are prescribed, within a viscoplastic framework, in terms of the following *effective slip rate*:

$$\dot{\gamma}_{\mathrm{eff}}^{(\beta)} = \sqrt{[\dot{\gamma}^{(\beta)}]^2 + [L_{\perp}\dot{\rho}_{\perp}^{(\beta)}]^2 + [L_{\odot}\dot{\rho}_{\odot}^{(\beta)}]^2 + [L_{\mathrm{M}}\dot{\rho}_{\mathrm{M}}^{(\beta)}]^2}\ \ \forall\beta \tag{5.40}$$

$$q^{(\beta)} = q_{\text{eff}}^{(\beta)} \frac{\dot{\gamma}^{(\beta)}}{\dot{\gamma}_{\text{eff}}^{(\beta)}}, \qquad \omega_I^{(\beta)} = L_I^2 q_{\text{eff}}^{(\beta)} \frac{\dot{\rho}_I^{(\beta)}}{\dot{\gamma}_{\text{eff}}^{(\beta)}} \text{ with } I = \perp, \odot, \text{M} \quad \forall \beta \tag{5.41}$$

where $L_\perp$, $L_\odot$, and L_M are *dissipative material length scales* and the effective stress, $q_{\text{eff}}^{(\beta)}$, which is work conjugate to the effective slip rate, is given by

$$q_{\text{eff}}^{(\beta)} = \sqrt{[q^{(\beta)}]^2 + \left[\frac{\omega_\perp^{(\beta)}}{L_\perp}\right]^2 + \left[\frac{\omega_\odot^{(\beta)}}{L_\odot}\right]^2 + \left[\frac{\omega_\text{M}^{(\beta)}}{L_\text{M}}\right]^2} \quad \forall \beta$$

The above prescription satisfies the second law of thermodynamics, as the dissipation rate turns out to be

$$\sum_\beta \left(q^{(\beta)} \dot{\gamma}^{(\beta)} + \omega_\perp^{(\beta)} \dot{\rho}_\perp^{(\beta)} + \omega_\odot^{(\beta)} \dot{\rho}_\odot^{(\beta)} + \omega_\text{M}^{(\beta)} \dot{\rho}_\text{M}^{(\beta)} \right) \equiv \sum_\beta q_{\text{eff}}^{(\beta)} \dot{\gamma}_{\text{eff}}^{(\beta)} > 0 \quad \forall \underline{\dot{\gamma}} \neq \underline{0}$$

By setting $L_\perp = L_\odot = L_\text{M} \equiv L$, the effective slip rate (5.40) simply becomes (Bardella 2006)

$$\dot{\gamma}_{\text{eff}}^{(\beta)} = \sqrt{[\dot{\gamma}^{(\beta)}]^2 + L^2 \dot{\gamma}_{,i}^{(\beta)} \dot{\gamma}_{,i}^{(\beta)}} \quad \forall \beta$$

By setting $L_\text{M} \equiv 0$, the higher-order modeling becomes purely based on dislocation densities (Gurtin et al. 2007; Gurtin and Ohno 2011; Gurtin and Reddy 2014). By neglecting the defect energy contribution, one recovers the model of Borg and Fleck (2007). Finally, the effective slip rate can be generalized by introducing a p-norm

$$\dot{\gamma}_{\text{eff}}^{(\beta)} = \left\{ |\dot{\gamma}^{(\beta)}|^p + |L_\perp \dot{\rho}_\perp^{(\beta)}|^p + |L_\odot \dot{\rho}_\odot^{(\beta)}|^p + |L_\text{M} \dot{\rho}_\text{M}^{(\beta)}|^p \right\}^{1/p} \quad \forall \beta$$

with $p > 0$ a material parameter usually set to a value in the range $[1, 2]$. The effect of this extension has been studied by Bardella and Giacomini (2008). In the context of phenomenological plasticity, it was earlier introduced by Fleck and Hutchinson (1997) and later discussed by Evans and Hutchinson (2009). Setting $p \neq 2$ should be strongly motivated by the need to model specific size-effects, as it may lead to significantly more difficult analyses.

The last constitutive law to be prescribed is that for $q_{\text{eff}}^{(\beta)}$. As most common in the literature, we assume that

$$q_{\text{eff}}^{(\beta)} = g^{(\beta)}(\underline{\hat{\gamma}}_{\text{eff}}) V(\dot{\gamma}_{\text{eff}}^{(\beta)}) \quad \forall \beta \quad \text{with} \quad \underline{\hat{\gamma}}_{\text{eff}} = \int \underline{\dot{\gamma}}_{\text{eff}} \, \text{d}t \tag{5.42}$$

where $V(\dot{\gamma}_{\text{eff}}^{(\beta)})$ is the function governing the rate dependence as defined in Eq. (5.31), and $g^{(\beta)}(\underline{\hat{\gamma}}_{\text{eff}})$ controls the self and latent isotropic hardening.

The Flow Rule

By substituting the constitutive laws into the modified Schmid law (5.37), we obtain the following system of partial differential equations (PDEs) in terms of kinematics:

$$\underbrace{g^{(\beta)} V(\dot{\gamma}^{(\beta)}_{\text{eff}}) \frac{\dot{\gamma}^{(\beta)}}{\dot{\gamma}^{(\beta)}_{\text{eff}}} - C^{(\beta)}_{ij} \left[g^{(\beta)} V(\dot{\gamma}^{(\beta)}_{\text{eff}}) \frac{\dot{\gamma}^{(\beta)}_{,j}}{\dot{\gamma}^{(\beta)}_{\text{eff}}} \right]_{,i}}_{\propto \text{ dissipative hardening}} + 2G\mu^{(\beta)}_{ij} \sum_{\kappa} \gamma^{(\kappa)} \mu^{(\kappa)}_{ij}$$

$$\underbrace{-N^{(\beta)}_{kil} \left(\frac{\mathrm{d}\mathcal{D}}{\mathrm{d}\alpha_{kl}} \right)_{,i}}_{\propto \text{ energetic backstress}} = 2G\mu^{(\beta)}_{ij} u_{i,j} \qquad \forall \beta$$

where G is the shear modulus and

$$C^{(\beta)}_{ij} = L^2_{\perp} s^{(\beta)}_i s^{(\beta)}_j + L^2_{\odot} l^{(\beta)}_i l^{(\beta)}_j + L^2_{\text{M}} m^{(\beta)}_i m^{(\beta)}_j$$

is a crystallographic tensor.

On the Higher-Order Boundary Conditions

If any of $L_{\perp}$, $L_{\odot}$, L_{M} is nonvanishing, the homogeneous higher-order boundary conditions are those specified in Eqs. (5.38) and (5.39) (Bardella 2006). Instead, if one neglects the higher-order dissipative stresses by setting $L_{\perp} = L_{\odot} = L_{\text{M}} = 0$, integration by parts of the internal virtual work (5.33) leads to (Gurtin and Needleman 2005)

$$\int_{\partial\Pi} \sum_{\beta} \xi^{(\beta)}_i \delta\gamma^{(\beta)} n_i \mathrm{d}A = \int_{\partial\Pi} \zeta_{ij} \underbrace{e_{jkl} \delta\gamma_{il} n_k}_{\delta\kappa_{ij}} \mathrm{d}A \qquad (5.43)$$

such that the kinematic higher-order boundary conditions have to be imposed on $\dot{\kappa}_{ij} = n_k \sum_{\beta} \dot{\gamma}^{(\beta)} N^{(\beta)}_{ikj}$ if the Schmid tensors $s^{(\beta)}_i m^{(\beta)}_j$ of the active slip systems span the space of deviatoric second-order tensors. Otherwise, the kinematic higher-order boundary conditions have to be imposed on $\dot{\underline{\gamma}}$.

Similarly, the microfree boundary conditions (5.38) may need to be substituted by $e_{jqk}\zeta_{iq}n_k - e_{pqk}\zeta_{pq}n_k\delta_{ij}/3 = 0$ on S^{dis}_T (Gurtin and Needleman 2005).

This has a relevant impact on the boundary value problem to be solved and on its FE implementation. This issue has been recently addressed by Panteghini and Bardella (2018) in the simpler context of Gurtin (2004) phenomenological SGP, later presented in section "Distortion Gradient Plasticity".

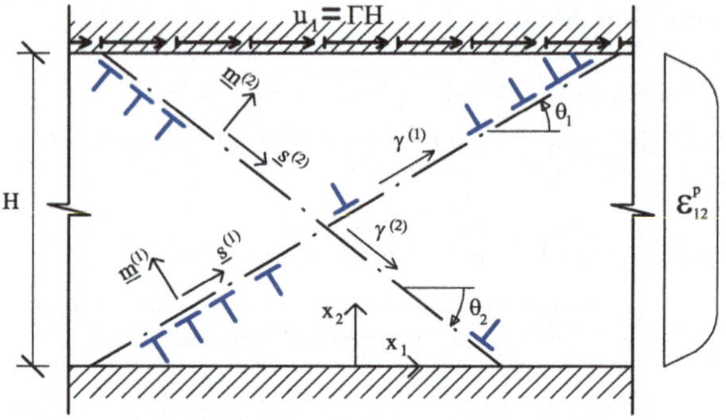

Fig. 5.5 Schematic of the simple shear benchmark

Size-Effects Describable by Strain Gradient Crystal Plasticity

The size-effects that can be described by the SGCP theory presented in this section have been initially demonstrated in the literature by focusing on the "simple shear benchmark," depicted in Fig. 5.5. This benchmark consists of a long strip (unbounded along the x_1 direction) of height H subjected to plane strain conditions and sheared between two bodies impenetrable to dislocations, such that the boundary value problem is spatially one-dimensional. The conventional boundary conditions are given by

$$u_1(x_2 = 0) = u_2(x_2 = 0) = u_2(x_2 = H) = 0\,, \quad u_1(x_2 = H) = \Gamma H \tag{5.44}$$

where Γ, which is referred to as the applied shear, is the average of $u_{1,2}(x_2)$ along the strip height.

We consider only slip systems with crystallographic vectors ($\boldsymbol{s}^{(\beta)}, \boldsymbol{m}^{(\beta)}$) in the shearing plane, implying that only edge dislocations (represented in blue in Fig. 5.5) are involved in this benchmark, which leads to $\varepsilon^p_{33} = 0$. The problem involves dislocation pileups at the boundaries where microhard boundary conditions hold, leading to vanishing plastic flow at the boundaries, as qualitatively illustrated in Fig. 5.5 for the plastic strain component ε^p_{12}.

Each slip system β is defined by its angle θ_β with respect to the x_1-axis, such that the crystallographic basis is given by

$$\boldsymbol{s}^{(\beta)} = [\cos\theta_\beta \quad \sin\theta_\beta \quad 0]^T\,, \quad \boldsymbol{m}^{(\beta)} = [-\sin\theta_\beta \quad \cos\theta_\beta \quad 0]^T\,, \quad \boldsymbol{l}^{(\beta)} = [0 \quad 0 \quad 1]^T$$

Hence, the in-plane components of the Schmid tensors are

$$\boldsymbol{s}^{(\beta)} \otimes \boldsymbol{m}^{(\beta)}\Big|_{\text{in-plane}} = \begin{bmatrix} -\sin(2\theta_\beta)/2 & \cos^2\theta_\beta \\ -\sin^2\theta_\beta & \sin(2\theta_\beta)/2 \end{bmatrix}$$

such that plastic strain, plastic spin, and Nye's tensor are expressed as

$$\boldsymbol{\varepsilon}^p\Big|_{\text{in-plane}} = \sum_\beta \frac{\gamma^{(\beta)}}{2}\begin{bmatrix} -\sin(2\theta_\beta) & \cos(2\theta_\beta) \\ \cos(2\theta_\beta) & \sin(2\theta_\beta) \end{bmatrix}$$

$$\boldsymbol{\vartheta}^p\Big|_{\text{in-plane}} = \sum_\beta \frac{\gamma^{(\beta)}}{2}\begin{bmatrix} 0 & 1 \\ -1 & 0 \end{bmatrix}$$

$$\boldsymbol{\alpha} = \sum_\beta \gamma^{(\beta)}_{,2}\begin{bmatrix} 0 & 0 & \sin(2\theta_\beta)/2 \\ 0 & 0 & \sin^2\theta_\beta \\ 0 & 0 & 0 \end{bmatrix}$$

If any of $L_\perp$, $L_\odot$, L_{M} is nonvanishing, the microhard boundary conditions read

$$\gamma^{(\beta)}(x_2 = 0) = \gamma^{(\beta)}(x_2 = H) = 0 \quad \forall\beta$$

otherwise, they become

$$\sum_\beta \gamma^{(\beta)}(x_2 = 0)\sin(2\theta_\beta) = \sum_\beta \gamma^{(\beta)}(x_2 = H)\sin(2\theta_\beta) = 0$$

and

$$\sum_\beta \gamma^{(\beta)}(x_2 = 0)\sin^2\theta_\beta = \sum_\beta \gamma^{(\beta)}(x_2 = H)\sin^2\theta_\beta = 0 \tag{5.45}$$

because the sole nonvanishing components of κ_{ij}, as defined in Eq. (5.43), are $\kappa_{13} = \sum_\beta \gamma^{(\beta)}\sin(2\theta_\beta)/2$ and $\kappa_{23} = \sum_\beta \gamma^{(\beta)}\sin^2\theta_\beta$.

Symmetric Double Slip

The case of "symmetric double slip," characterized by couples of slip systems symmetrically oriented with respect to the x_1-axis, is particularly simple to deal with, as the sole relevant plasticity fields become

$$\varepsilon^p_{12} = \sum_\beta \cos(2\theta_\beta)\gamma^{(\beta)}, \quad \vartheta^p_{12} = \sum_\beta \gamma^{(\beta)}, \quad \alpha_{23} = 2\sum_\beta \sin^2\theta_\beta\gamma^{(\beta)}_{,2}$$

where now β indicates a couple of symmetrically oriented slip systems. The sole nonvanishing Cauchy stress component is σ_{12} which is spatially uniform.

"Perfect Rate-Independent Plasticity" with Quadratic Defect Energy in the Absence of Higher-Order Dissipative Stresses

This case of *symmetric double slip* can be solved analytically (Bardella 2006; Bardella and Giacomini 2008). It is an interesting case as it reveals the main effect of prescribing a quadratic defect energy and it provides an instructive example of non-uniqueness of solution even in the framework of higher-order SGCP.

For the defect energy, we adopt the following quadratic form, dependent on a single *energetic length scale* ℓ and modulated by the shear modulus G:

$$\mathcal{D}(\boldsymbol{\alpha}) = \frac{1}{2} G \ell^2 \alpha_{ij} \alpha_{ij}$$

In "perfect rate-independent plasticity," the prescription (5.42) is simply substituted by

$$|q^{(\beta)}| = g_0 \quad \forall \beta$$

if plasticity occurs, where g_0 is a material constant representing the slip resistance and $q_{\text{eff}}^{(\beta)} \equiv |q^{(\beta)}|$ given the assumption $L_{\perp} = L_{\odot} = L_{\text{M}} = 0$.

The balance equations are

$$\sigma_{12} = \text{constant} \quad \text{and} \quad q^{(\beta)} = \cos(2\theta_\beta)\sigma_{12} + \sin^2\theta_\beta\, \zeta_{23,2} \quad \forall \beta \tag{5.46}$$

that have to be solved, subject to the boundary conditions (5.44) and (5.45), together with the following equations, involving compatibility and constitutive conditions:

$$u_{1,2} - 2\sum_\beta \gamma^{(\beta)} \cos(2\theta_\beta) = \frac{\sigma_{12}}{G}$$

$$2\sum_\beta \gamma_{,2}^{(\beta)} \sin^2\theta_\beta = \frac{\zeta_{23}}{G\ell^2} \tag{5.47}$$

as well as the KKT and consistency conditions (5.29) and (5.30) with the "yield function" defined from (5.46) as $\varphi^{(\beta)} = \cos(2\theta_\beta)\sigma_{12} + \sin^2\theta_\beta\, \zeta_{23,2} - q^{(\beta)}$.

Henceforth, for the sake of simplicity, we assume that (i) $\Gamma \geq 0$ increases monotonically and (ii) $0 \leq \theta_1 < \theta_2 < \theta_3 < \cdots < \theta_B \leq \pi/4$ with B denoting the number of couples of slip systems that can be activated.

Let us first focus on conventional crystal plasticity. If one neglects higher-order effects by setting $\ell = 0$, one has $q^{(\beta)} = \cos(2\theta_\beta)\sigma_{12}$ $\forall \beta$ and the solution is linear elastic until the applied strain assumes the value

$$\Gamma = \Gamma_{(1)} \equiv \frac{g_0}{G\cos(2\theta_1)},$$

corresponding to the limit stress $q^{(1)} = g_0$ on the less inclined couple of slip systems. After this stage, the elastic strain remains constant even if Γ increases, whereas the plastic slip $\gamma^{(1)}$ may assume any positive distribution whose average over the strip height accommodates the increment in the applied shear with respect to $\Gamma_{(1)}$. Among these distributions, one may select a solution satisfying the microhard boundary conditions, as, for instance, the parabolic solution obtainable as the limit for $\ell \to 0$ of the solution for $B = 1$ and $\ell > 0$ (Bardella 2006) according to

$$\gamma^{(1)} = 3\frac{\Gamma \cos(2\theta_1) - g_0/G}{\cos^2(2\theta_1)}\left(1 - \frac{x_2}{H}\right)\frac{x_2}{H} \tag{5.48}$$

Here, the solution of the SGCP problem, analyzed next, acts as a selector among the infinite solutions of the conventional problem.

For $\ell > 0$, the behavior is still linear elastic until $\Gamma = \Gamma_{(1)}$. Then, plasticity develops on slip system 1 and we obtain $\gamma^{(1)}$ by integration of

$$g_0 = \cos(2\theta_1)\sigma_{12} + 2G\ell^2 \sin^4\theta_1 \gamma^{(1)}_{,22},$$

obtained by combining (5.46) and (5.47). The solution, for $\Gamma \geq \Gamma_{(1)}$ and before a second couple of slip systems gets activated, is given by

$$\sigma_{12} = \frac{\Gamma G + K_1 g_0}{1 + K_1 \cos(2\theta_1)} \tag{5.49}$$

$$\gamma^{(1)} = \frac{\Gamma \cos(2\theta_1) - g_0/G}{4 \sin^4\theta_1 [1 + K_1 \cos(2\theta_1)]}\left(\frac{H}{\ell}\right)^2\left(1 - \frac{x_2}{H}\right)\frac{x_2}{H} \tag{5.50}$$

where

$$K_1 = \frac{\cos(2\theta_1)}{12 \sin^4\theta_1}\left(\frac{H}{\ell}\right)^2 \tag{5.51}$$

The limit for $\ell \to 0$ of Eqs. (5.50) and (5.51) leads to Eq. (5.48). The quadratic profile for $\gamma^{(1)}$ established by Eq. (5.50) confirms the results of Bittencourt et al. (2003). More importantly, inspection of Eqs. (5.49) and (5.51) clearly shows a size-effect in the macroscopic response in terms of shear stress σ_{12} as a function of the applied shear Γ. In particular, the derivative

$$\frac{d\sigma_{12}}{d\Gamma} = \frac{G}{1 + \dfrac{(H/\ell)^2}{12 \sin^4\theta_1}}$$

demonstrates that the size-effect consists of an increase of strain hardening with diminishing size, whereas the yield point, $\sigma_{12} = g_0/\cos(2\theta_1)$, is in this case unaffected by the size (i.e., there is no *strengthening*).

Moreover, ℓ governs the thickness of the boundary layers, as can be found from Eqs. (5.50) and (5.51), by evaluating $\mathrm{d}\gamma^{(1)}/\mathrm{d}x_2|_{x_2=0,H}$ and assuming that the slope of the plastic slip at the boundary is proportional to this thickness. This effect of ℓ in a quadratic defect energy is clearly seen by including some positive self hardening (Anand et al. 2005; Bardella 2006), here skipped for brevity.

Contrary to the case of conventional crystal plasticity, when $\ell > 0$, after plasticity occurs, σ_{12} may increase because of the defect stress contribution in the higher-order balance Eq. (5.46) on the slip plane 1. This behavior holds until Γ reaches the value $\Gamma_{(\delta)}$ corresponding to $\sigma_{12} = (g_0 - \zeta_{23,2} \sin^2\theta_\delta)/\cos(2\theta_\delta)$ such that $q^{(\delta)} = \pm g_0$, thus activating the system δ (note that it is not necessarily $\delta = 2$, as demonstrated below). Now, the plastic slips are determined by solving

$$\begin{cases} g_0 = \cos(2\theta_1)\sigma_{12} + 2G\ell^2 \sin^2\theta_1 \big(\gamma^{(1)}_{,22} \sin^2\theta_1 + \gamma^{(\delta)}_{,22} \sin^2\theta_\delta\big) \\ \pm g_0 = \cos(2\theta_\delta)\sigma_{12} + 2G\ell^2 \sin^2\theta_\delta \big(\gamma^{(1)}_{,22} \sin^2\theta_1 + \gamma^{(\delta)}_{,22} \sin^2\theta_\delta\big) \end{cases} \tag{5.52}$$

Note that the quantity

$$K_2 = 2G\ell^2 \big(\gamma^{(1)}_{,22} \sin^2\theta_1 + \gamma^{(\delta)}_{,22} \sin^2\theta_\delta\big) \tag{5.53}$$

must be constant, such that the solution of system (5.52) depends only on the two unknowns K_2 and σ_{12} that result

$$\sigma_{12} = \frac{g_0}{K_3}(\sin^2\theta_\delta \mp \sin^2\theta_1)$$

$$K_2 = \frac{g_0}{K_3}[\pm \cos(2\theta_1) - \cos(2\theta_\delta)] \tag{5.54}$$

with

$$K_3 = \cos(2\theta_1)\sin^2\theta_\delta - \cos(2\theta_\delta)\sin^2\theta_1 > 0$$

Moreover, by combining Eqs. (5.53) and (5.54) with the microhard boundary conditions (5.45) one finds

$$\gamma^{(1)}(x_2)\sin^2\theta_1 + \gamma^{(\delta)}(x_2)\sin^2\theta_\delta = \left(\frac{H}{\ell}\right)^2 \frac{g_0}{4GK_3}[\cos(2\theta_\delta) \mp \cos(2\theta_1)]\Big(1 - \frac{x_2}{H}\Big)\frac{x_2}{H} \tag{5.55}$$

Since $\gamma^{(1)} > 0$ at incipient plasticity on the system δ (i.e., when $\Gamma = \Gamma_{(\delta)}$), relation (5.55) must hold with the plus sign; because of relations (5.47) and (5.52), this requires $\gamma^{(\delta)} < 0$. We note that the combination $\gamma^{(1)}(x_2) > 0$ and $\gamma^{(\delta)}(x_2) < 0$ allows nonvanishing values of the slips at the boundary ($x_2 = 0$ and $x_2 = H$).

We finally obtain

$$\sigma_{12} = \frac{g_0}{K_3}(\sin^2\theta_\delta + \sin^2\theta_1) \tag{5.56}$$

and note that Eq. (5.55) allows $\gamma^{(1)}$ to be expressed in terms of $\gamma^{(\delta)}$.

The remaining condition to be imposed is the conventional boundary condition (5.44), i.e.,

$$\Gamma H = \int_0^H u_{1,2}(x_2)\,\mathrm{d}x_2$$

which, by inspection of Eqs. (5.47), (5.55), and (5.56), leads to the conclusion that $\gamma^{(\delta)}$ *is any negative distribution* meeting the requirement

$$\Gamma = \Gamma_{(\delta)} - \frac{2K_3}{\sin^2\theta_1}\frac{1}{H}\int_0^H \gamma^{(\delta)}(x_2)\,\mathrm{d}x_2$$

In fact, relation (5.55) constrains $\gamma^{(1)}$ to evolve along with $\gamma^{(\delta)}$ in such a way that the defect stress remains constant, i.e., independent of any $\Gamma \geq \Gamma_{(\delta)}$, whereas the maximum allowed Cauchy stress value is given by (5.56). Hence, no more slip systems can be activated.

Finally, we determine $\Gamma_{(\delta)}$ by looking at the higher-order balance equation on the glide δ at incipient plasticity, that is, when $\gamma^{(\delta)}$ is very close to zero (though negative) and $|q^{(\delta)}|$ has just reached the value g_0 given by

$$-g_0 = \cos(2\theta_\delta)\sigma_{12} + 2G\ell^2\gamma^{(1)}_{,22}\sin^2\theta_1\sin^2\theta_\delta$$

This relation, combined with Eqs. (5.49) and (5.50), provides

$$\Gamma_{(\delta)} = \frac{g_0}{GK_3}\Bigg[\sin^2\theta_1 + \sin^2\theta_\delta + \left(\frac{H}{\ell}\right)^2 \frac{\cos(2\theta_1)}{12\sin^2\theta_1}[\cos(2\theta_1) + \cos(2\theta_\delta)]\Bigg] \tag{5.57}$$

By minimizing the right-hand side of Eq. (5.57) with respect to $\theta_\delta \in \{\theta_2, \theta_3, \ldots, \theta_B\}$, we can precisely determine the system δ among those allowed by the chosen crystallography.

The foregoing analysis demonstrates that non-uniqueness of solution may occur even for a well-established higher-order strain gradient crystal plasticity. Thus, when higher-order terms consist only of the defect energy we expect numerical issues in rate-independent perfect plasticity. Accounting for higher-order dissipation (i.e., setting any $L_I > 0$, $I = \perp, \odot, \mathrm{M}$, in the present theory) provides a regularization addressing this issue. By referring to plastic strain localization (that may be a numerical outcome due to non-uniqueness of solution), this result of Bardella and Giacomini (2008) provides an explanation for the numerical findings of Bittencourt et al. (2003), who implemented Gurtin (2002) SGCP and applied it to a fiber composite benchmark. Within this benchmark, a similar localization phenomenon has been numerically observed by Poh and Peerlings (2016), who implemented the phenomenological SGP theory of Gurtin (2004) (here presented in section "Distortion Gradient Plasticity"), and later through the H(curl) FE analysis of Panteghini and Bardella (2018) of the same theory applied to a slightly different benchmark. The fundamental common feature of all these analyses is that the higher-order contribution depends exclusively on Nye's tensor.

Size-Effects Ensuing from Higher-Order Dissipation and Interaction Among Slip Systems

As demonstrated in Bardella (2006), the higher-order dissipation controls the increase of the yield point with diminishing size, through the dissipative length scales L_I $(I = \perp, \odot, \mathrm{M})$, without much affecting the strain hardening nor boundary layer thicknesses. This is the size-effect referred to as *strengthening*. Earlier results on this are due to Anand et al. (2005) who investigated a one-dimensional strain gradient plasticity theory, whereas Chiricotto et al. (2016) have more recently unveiled the mathematical reasons for the prediction of strengthening. Very recently, on the basis of the same one-dimensional benchmark, Willis (2019) has explored a more general higher-order dissipative law, though by referring to phenomenological SGP and with focus on a specific phenomenon referred to as "elastic gap," here illustrated in the torsion benchmark of section "Distortion Gradient Plasticity".

The interaction between the crystallography and the size-effects describable by energetic and dissipative higher-order contributions can be appreciated by studying a material admitting some (isotropic) self hardening, such that many slip systems can get active without facing any limit load nor non-uniqueness of solution.

Finally, an attempt to predict the results of discrete dislocation dynamics simulations through SGCP (Bardella et al. 2013) has shown the need to include a specific size dependence in the constitutive laws governing the latent hardening in SGCP.

Phenomenological Strain Gradient Plasticity

In contrast to *crystal* theories, *phenomenological* plasticity theories disregard the details of the crystallographic structure, such that the fundamental plasticity variable consists merely of the plastic distortion u^p_{ij}.

Under fairly general assumptions, the internal and external virtual works (to be plugged into the *generalized* PVW (5.20)) may be written, in any subregion $\Pi \subset \Omega$ and in the absence of body forces, as

$$\mathcal{W}_i = \int_\Pi \left[\sigma_{ij}(\delta\varepsilon_{ij} - \delta\varepsilon^p_{ij}) + \varsigma_{ij}\delta u^p_{ij} + S_{ijk}\delta u^p_{ij,k}\right]\mathrm{d}V \tag{5.58}$$

$$\mathcal{W}_e = \int_{\partial\Pi} \left[T_i\delta u_i + t_{ij}\delta u^p_{ij}\right]\mathrm{d}A \tag{5.59}$$

where t_{ij} are the higher-order tractions.

From Eqs. (5.20), (5.58), and (5.59), depending on the constitutive assumptions for the non-conventional stresses ς_{ij} and S_{ijk}, work conjugate to the plastic distortion and its gradient, respectively, one can obtain most of the higher-order SGP theories proposed in the literature. For isochoric plastic flow, we take $\varsigma_{ii} = S_{iik} = 0$.

The generalized PVW (through integration by parts, application of the divergence theorem, and standard variational arguments) leads to the conventional balance Eqs. (5.26)–(5.25) along with the higher-order balance equation

$$\sigma_{ij}^{\mathrm{d}} - \varsigma_{ij} + S_{ijk,k} = 0 \quad \text{in } \Omega \tag{5.60}$$

and the microfree boundary conditions (prescribing $t_{ij} = 0$ on S_T^{dis})

$$S_{ijk} n_k = 0 \quad \text{on } S_T^{\mathrm{dis}}$$

Here and henceforth, σ_{ij}^{d} is the deviatoric part of the Cauchy stress tensor.

Though strain gradient plasticity has a relatively general meaning, in the following we distinguish between distortion gradient plasticity, in which plastic spin contributes both to the free energy density and, eventually, to the dissipation, and strain gradient plasticity, in which there are no stresses associated with the plastic spin, which is ensured by the assumption $S_{ijk} = S_{jik}$.

Distortion Gradient Plasticity

Here, we basically refer to the theory proposed by Gurtin (2004). It can be obtained by assuming, in the internal virtual work (5.58), that

$$S_{ijk} = S_{ijk}^{\mathrm{def}} + T_{ijk}^{(\varepsilon)} \tag{5.61}$$

where

$$S_{ijk}^{\mathrm{def}} = e_{kjh}\zeta_{ih} - \frac{1}{3}\delta_{ij} e_{kph}\zeta_{ph} \quad \text{and} \quad T_{ijk}^{(\varepsilon)} = T_{jik}^{(\varepsilon)} \tag{5.62}$$

in which ζ_{ij} is the defect stress as defined in Eq. (5.34), thus being a recoverable higher-order stress work conjugate to Nye's tensor, whereas $T_{ijk}^{(\varepsilon)}$ is a *dissipative* higher-order stress work conjugate to the *rate* of the plastic strain gradient.

Hence, the internal virtual work (5.58) specializes to

$$\mathcal{W}_i = \int_{\Pi} \Big[\underbrace{\sigma_{ij}(\delta\varepsilon_{ij} - \delta\varepsilon_{ij}^p) + \zeta_{ij}\delta\alpha_{ij}}_{\text{energetic}} + \underbrace{\varsigma_{ij}\delta u_{ij}^p + T_{ijk}^{(\varepsilon)}\delta\varepsilon_{ij,k}^p}_{\text{dissipative}} \Big] \mathrm{d}V \tag{5.63}$$

in which the assumption that ς_{ij} is purely dissipative implies neglecting conventional kinematic hardening.

From the *generalized* PVW (5.20), one obtains the conventional balance conditions (5.26) and (5.25) along with the following higher-order balance equations:

$$\sigma^{\mathrm{d}}_{ij} - \left[\frac{1}{2}\left(e_{jkl}\zeta_{il,k} + e_{ikl}\zeta_{jl,k}\right) - \frac{1}{3}\delta_{ij}e_{mkl}\zeta_{ml,k}\right] = \rho_{ij} - T^{(\varepsilon)}_{ijk,k} \quad \text{in} \quad \Omega \tag{5.64}$$

$$\frac{1}{2}\left(e_{jkl}\zeta_{il,k} - e_{ikl}\zeta_{jl,k}\right) = \omega_{ij} \quad \text{in} \quad \Omega \tag{5.65}$$

where

$$\rho_{ij} = \frac{1}{2}\left(\varsigma_{ij} + \varsigma_{ji}\right) \quad \text{and} \quad \omega_{ij} = \frac{1}{2}\left(\varsigma_{ij} - \varsigma_{ji}\right)$$

and *microfree* boundary conditions

$$\left[\frac{1}{2}\left(e_{jlk}\zeta_{il} + e_{ilk}\zeta_{jl}\right) - \frac{1}{3}\delta_{ij}e_{mlk}\zeta_{ml}\right]n_k + T^{(\varepsilon)}_{ijk}n_k = 0 \quad \text{on} \quad S^{\mathrm{dis}}_T \tag{5.66}$$

$$\left(e_{jlk}\zeta_{il} - e_{ilk}\zeta_{jl}\right)n_k = 0 \quad \text{on} \quad S^{\mathrm{dis}}_T \tag{5.67}$$

Equations (5.64) and (5.65) can also be obtained from (5.60) under the assumptions (5.61) and (5.62). In fact, while Eq. (5.60) includes eight PDEs obtained by imposing the independence of the variation of the whole plastic distortion, Eq. (5.64) includes five scalar PDEs obtained by imposing the independence of the variation of the plastic strain, whereas Eq. (5.65) includes three scalar PDEs obtained by imposing the independence of the variation of the plastic spin. This subdivision in two distinct higher-order balance equations is convenient in view of more specific constitutive prescriptions. The same holds for the microfree boundary conditions (5.66) and (5.67).

By integrating the internal virtual work (5.63) by parts, it can be recognized that the microhard boundary conditions depend on whether the dissipative higher-order stress $T^{(\varepsilon)}_{ijk}$ is included or not in the model. We note that it can be excluded just by setting to zero the material length scale entering its prescription in terms of the rate of the plastic strain gradient (as defined in the following). Hence, if $T^{(\varepsilon)}_{ijk}$ enters the model, the microhard boundary conditions read

$$\dot{\varepsilon}^{p}_{ij} = 0 \quad \text{and} \quad e_{jlk}\vartheta^{p}_{il}n_k = 0 \quad \text{on} \quad S^{\mathrm{dis}}_U$$

Otherwise, if $T^{(\varepsilon)}_{ijk}$ is constitutively set to zero, the microhard boundary conditions are given by

$$e_{jlk}u^{p}_{il}n_k = 0 \quad \text{on} \quad S^{\mathrm{dis}}_U \tag{5.68}$$

This is an important issue that has relevant implications on the solution of boundary value problems and on their FE implementation. In the plane strain case, the theory including $T^{(\varepsilon)}_{ijk}$ has been implemented in two different FE frameworks by Martínez-

Pañeda et al. (2016) and Panteghini and Bardella (2016), who have studied the simple shear benchmark, the bending of thin foils, and a simplified indentation benchmark. Both implementations share the use of all plastic distortion components as nodal degrees of freedom, as this is possible because in plane strain only one plastic spin component is nonvanishing. The theory constitutively neglecting $T^{(\varepsilon)}_{ijk}$ requires special H(curl) FEs, developed by Panteghini and Bardella (2018) still in the plane strain case, allowing discontinuities in the plastic distortion components, according to Eq. (5.68). Here, we apply Gurtin (2004) theory by focusing on the torsion problem, in which the two nonvanishing components of the plastic distortion (that can be written in terms of $\varepsilon^p(r)$ and $\vartheta^p(r)$ as in (5.12)) can be chosen as nodal degrees of freedom in a spatially one-dimensional FE.

Constitutive Relations: The Dissipative Behavior and the Dissipative Length Scales

The dissipative stresses are prescribed, within a viscoplastic framework, in terms of the following *effective plastic flow rate*:

$$\dot{E}^p = \sqrt{\frac{2}{3}\dot{\varepsilon}^p_{ij}\dot{\varepsilon}^p_{ij} + \chi\dot{\vartheta}^p_{ij}\dot{\vartheta}^p_{ij} + \frac{2}{3}L^2\dot{\varepsilon}^p_{ij,k}\dot{\varepsilon}^p_{ij,k}} \tag{5.69}$$

$$\rho_{ij} = \frac{2}{3}\frac{\Sigma}{\dot{E}^p}\dot{\varepsilon}^p_{ij}\,, \quad \omega_{ij} = \chi\frac{\Sigma}{\dot{E}^p}\dot{\vartheta}^p_{ij}\,, \quad T^{(\varepsilon)}_{ijk} = \frac{2}{3}L^2\frac{\Sigma}{\dot{E}^p}\dot{\varepsilon}^p_{ij,k} \tag{5.70}$$

where χ is a material parameter controlling the dissipation due to the plastic spin, L is a *dissipative length scale*, and Σ is the effective stress which turns out to be work conjugate to the effective plastic flow rate

$$\Sigma = \sqrt{\frac{3}{2}\rho_{ij}\rho_{ij} + \frac{1}{\chi}\omega_{ij}\omega_{ij} + \frac{3}{2L^2}T^{(\varepsilon)}_{ijk}T^{(\varepsilon)}_{ijk}}$$

This prescription is thermodynamically consistent as the dissipation rate results in

$$\rho_{ij}\dot{\varepsilon}^p_{ij} + \omega_{ij}\dot{\vartheta}^p_{ij} + T^{(\varepsilon)}_{ijk}\dot{\varepsilon}^p_{ij,k} \equiv \Sigma\dot{E}^p > 0 \ \ \forall\dot{u}^p_{ij} \neq 0$$

The last constitutive law to be prescribed is that for Σ. As most common in the literature, we assume

$$\Sigma(\dot{E}^p, E^p) = \sigma_Y(E^p)V(\dot{E}^p) \quad \text{with} \quad E^p = \int \dot{E}^p \mathrm{d}t$$

where $V(\dot{E}^p)$ is the function governing the rate dependence and $\sigma_Y(E^p)$ controls the isotropic hardening. Here, for $V(\dot{E}^p)$ we adopt the viscoplastic function (5.31) in which, because of the phenomenological context, we change the symbol for the rate sensitivity, thus substituting $\dot{\gamma}_0$ with $\dot{\varepsilon}_0$.

For the isotropic hardening, we assume the following form:

$$\sigma_Y(E^p) = \sigma_0 + h \cdot (E^p)^{n_h} \tag{5.71}$$

where σ_0 is the initial yield stress and h and n_h are hardening parameters.

Minimum Principles

By postulating the existence of a dissipation potential defined as

$$\mathcal{V}(\dot{E}^p, E^p) = \int_0^{\dot{E}^p} \Sigma(\dot{E}^{p\prime}, E^p)\mathrm{d}\dot{E}^{p\prime} \tag{5.72}$$

minimum principles equivalent to the balance equations hold under the assumption that both the dissipation potential and the defect energy are convex functions. Such minimum principles are surely useful in computation and, for the theory of Gurtin (2004), have been in fact obtained for this purpose by Martínez-Pañeda et al. (2016) as an extension of the minimum principles developed by Fleck and Willis for SGP whose primal higher-order variable is either the gradient of the equivalent von Mises plastic strain (Fleck and Willis 2009a) or the plastic strain gradient (Fleck and Willis 2009b).

Under the kinematic constraints (5.2), (5.3), and (5.69), the field u^p_{ij} minimizing the functional

$$\mathcal{H}(\boldsymbol{u}^p) = \int_\Omega \left[\mathcal{V}(\dot{E}^p, E^p) + \zeta_{ij}\dot{\alpha}_{ij} - \sigma_{ij}\dot{\varepsilon}^p_{ij}\right]\mathrm{d}V$$

fulfills the higher-order balance Eqs. (5.64) and (5.65).

Additionally, for a given $\dot{\varepsilon}^p_{ij}$, the conventional balance Eqs. (5.26) and (5.25) are satisfied by any kinematically admissible field $\dot{u}_i$ minimizing the functional

$$\mathcal{J}(\dot{\boldsymbol{u}}) = \frac{1}{2}\int_\Omega (\dot{u}_{i,j} - \dot{\varepsilon}^p_{ij})L_{ijkl}(\dot{u}_{k,l} - \dot{\varepsilon}^p_{kl})\mathrm{d}V - \int_{S_T} T_i\dot{u}_i\mathrm{d}A$$

where $L_{ijkl} = L_{jikl} = L_{ijlk}$ (see Eq. (5.28)).

The Torsion of Thin Wires

The Nye tensor form (5.13) for this benchmark strongly suggests to express the defect energy as a function of the two invariants (5.14) and (5.15), $\mathcal{D}(\alpha_{ii}, |\alpha^{\mathrm{d}}|)$, such that the defect stress, in general, results as

$$\zeta_{ij} = \frac{\partial \mathcal{D}}{\partial \alpha_{kk}} \frac{\alpha_{ll}}{|\alpha_{mm}|} \delta_{ij} + \frac{\partial \mathcal{D}}{\partial |\alpha^{\mathrm{d}}|} \frac{\alpha^{\mathrm{d}}_{ij}}{|\alpha^{\mathrm{d}}|} \tag{5.73}$$

A possible choice of quadratic defect energy is

$$\mathcal{D} = \frac{1}{2} G(\ell_1 \alpha_{ii})^2 + \frac{1}{2} G(\ell_2 |\alpha^{\mathrm{d}}|)^2 \tag{5.74}$$

which depends on the two material length scales ℓ_1 and ℓ_2, and such that the defect stress (5.73) becomes

$$\zeta_{ij} = G\ell_1^2 \alpha_{kk} \delta_{ij} + G\ell_2^2 \alpha^{\mathrm{d}}_{ij}$$

We further consider a logarithmic defect energy, firstly adopted by Forest and Guéninchault (2013) on the basis of the studies of Groma et al. (2003, 2007) on the continuum description of dislocation dynamics. Here, we consider the following regularized logarithmic form (a similar regularization has been proposed by Svendsen and Bargmann 2010 within the context of crystal plasticity)

$$\mathcal{D} = G\ell_1 |\alpha_{ii}| \ln\left(\frac{\ell_1 |\alpha_{kk}|}{\alpha_1} + 1\right) + G\ell_2 |\alpha^{\mathrm{d}}| \ln\left(\frac{\ell_2 |\alpha^{\mathrm{d}}|}{\alpha_2} + 1\right) \tag{5.75}$$

which involves two further material parameters α_1 and α_2, with respect to the quadratic form (5.74). In this case, the defect stress (5.73) is

$$\zeta_{ij} = G\ell_1 |\alpha_{kk}| \left[\ln\left(\frac{\ell_1 |\alpha_{ll}|}{\alpha_1} + 1\right) + \frac{1}{1 + \dfrac{\alpha_1}{\ell_1 |\alpha_{mm}|}} \right] \frac{\alpha_{nn}}{|\alpha_{pp}|} \delta_{ij}$$
$$+ G\ell_2 |\alpha^{\mathrm{d}}| \left[\ln\left(\frac{\ell_2 |\alpha^{\mathrm{d}}|}{\alpha_2} + 1\right) + \frac{1}{1 + \dfrac{\alpha_2}{\ell_2 |\alpha^{\mathrm{d}}|}} \right] \frac{\alpha^{\mathrm{d}}_{ij}}{|\alpha^{\mathrm{d}}|}$$

The kinematics illustrated in section "The Kinematics of the Torsion of Thin Wires" leads to the following nonvanishing stress components:

$$\sigma \equiv \sigma_{\theta z}\,, \quad \zeta_{rr}\,, \quad \zeta_{\theta\theta}\,, \quad \zeta_{zz}\,, \quad \rho \equiv \rho_{\theta z}\,, \quad \omega \equiv \omega_{\theta z} = -\omega_{z\theta}\,, \quad T \equiv T^{(\varepsilon)}_{\theta zr} = T^{(\varepsilon)}_{z\theta r}$$

with the torque Q resulting from the applied twist κ under microfree boundary conditions given as

$$Q = 2\pi \int_0^a \sigma\, r^2 \mathrm{d}r$$

where a denotes the cross-section radius.

In the wire center dislocations pile up, such that we apply microhard boundary conditions

$$\varepsilon^p(0) = 0\,, \qquad \vartheta^p(0) = 0$$

We develop the FE algorithm by starting from the PVW, which must hold for any kinematically admissible independent variations $\delta\varepsilon^p$ and $\delta\vartheta^p$, which are the sole nodal degrees of freedom, as, in this problem, the displacement field is given by Eq. (5.11). Hence, the weak form of the balance corresponds to the higher-order balance equations and it may be stated as

$$\int_0^a \Big[(-2\sigma r - \zeta_{rr} + \zeta_{zz} + 2\rho r)\delta\varepsilon^p + (-\zeta_{\theta\theta} + \zeta_{zz} + 2T)r\delta(\varepsilon^p)'\Big]\mathrm{d}r = 0 \tag{5.76}$$

$$\int_0^a \Big[(-\zeta_{rr} - \zeta_{zz} + 2\omega r)\delta\vartheta^p + (-\zeta_{\theta\theta} - \zeta_{zz})r\delta(\vartheta^p)'\Big]\mathrm{d}r = 0 \tag{5.77}$$

With reference to the time discretization, we adopt the symbols ξ_t and $\Delta\xi$ for the nodal variable ξ (equal to either ε^p or ϑ^p) evaluated at the beginning of the time step and its increment within the time step, respectively. Thus, within each time step, ξ_t is known, contrary to $\Delta\xi$. We employ the backward Euler time integration scheme, such that we compute the relevant quantities at the end of the time step according to

$$\xi = \xi_t + \Delta\xi$$

whose rate is expressed in terms of the time step amplitude Δt as $\dot{\xi} = \Delta\xi/\Delta t$.

The FE spatial discretization relies on the interpolation functions $N_\xi^{(i)}(s)$, assuming value 1 at the node i and 0 at the other nodes; they are expressed in terms of the intrinsic coordinate $s \in [-1, 1]$ on the parent element. Moreover, we limit our numerical analysis to isoparametric FEs, such that $N_{\varepsilon^p}^{(i)}(s) = N_{\vartheta^p}^{(i)}(s) = N^{(i)}(s)$ and

$$r(s) = \sum_i r^{(i)} N^{(i)}(s)$$

with $r^{(i)}$ the nodal values of the radial coordinate. Therefore, we have

$$\xi_t(s) = \sum_i \xi_t^{(i)} N_\xi^{(i)}(s)\,, \qquad \Delta\xi(s) = \sum_i \Delta\xi^{(i)} N_\xi^{(i)}(s)$$

where $\xi_t^{(i)}$ and $\Delta\xi^{(i)}$ are nodal values.

The left-hand sides of the balance Eqs. (5.76) and (5.77) are implemented by integrating over a single FE, thus obtaining, $\forall\delta\varepsilon^p, \delta\vartheta^p$

$$\int_{-1}^1 \left[(-2\sigma r - \zeta_{rr} + \zeta_{zz} + 2\rho r)\, J N^{(i)} + (\zeta_{zz} - \zeta_{\theta\theta} + 2T)\, r\frac{\mathrm{d}N^{(i)}}{\mathrm{d}s}\right]\mathrm{d}s \tag{5.78}$$

$$\int_{-1}^{1}\left[(2\omega r-\zeta_{rr}-\zeta_{zz})\,J N^{(i)}-(\zeta_{zz}+\zeta_{\theta\theta})\,r\frac{\mathrm{d}N^{(i)}}{\mathrm{d}s}\right]\mathrm{d}s \tag{5.79}$$

Here and henceforth, $J = \mathrm{d}r/\mathrm{d}s$. The FE code checks whether the convergence is achieved or not on the basis of Eqs. (5.78) and (5.79). By differentiating the opposites of these expressions with respect to the incremental nodal variables, one obtains the consistent FE tangent stiffness, which is needed to guarantee second-order convergence of the FE algorithm (Ortiz and Popov 1985). The explicit expression of this operator is provided in Bardella and Panteghini (2015).

Henceforth, all the results refer to linear (two-noded) elements with reduced integration (i.e., they have a single Gauss point).

The convergence of the results has been checked as explained in Bardella and Panteghini (2015). In particular, in the case of quadratic defect energy in the absence of higher-order dissipation, a regular mesh of 100 FEs provides results in perfect agreement with the analytical ones obtained by Chiricotto et al. (2012). Such a mesh has been employed in all the analyses, if not otherwise specified in the following.

All the results presented next refer to shear modulus $G = 45$ GPa and almost rate independence, the latter obtained by setting $P = 0.05$ and an appropriately small value of $\dot{\varepsilon}_0$ (see Eq. (5.31)). The isotropic hardening parameters are $h = 150$ MPa and $n_h = 0.37$ (see Eq. (5.71)).

We first consider the response under monotonic loading and focus on the effect of the length scales ℓ_1 and ℓ_2 entering the quadratic defect energy (5.74). We set $L = 0$, $\sigma_0 = 55$ MPa, $\dot{\varepsilon}_0 = 1 \cdot 10^{-5}$ s^{-1}, and $\chi = 2/3$.

Figure 5.6 shows that the length scale ℓ_1, governing the influence of α_{ii}, controls the strain hardening increase with diminishing size, whereas Fig. 5.7, which is concerned with the length scale ℓ_2, governing the influence of $|\alpha^{\mathrm{d}}|$, displays a slightly different size-effect that actually includes some strengthening. This behavior is usually not observed in phenomenological plasticity if the defect energy is a plain quadratic function of its independent variable (e.g., $\mathcal{D} \propto \alpha_{ij}\alpha_{ij}$ by adopting Nye's tensor).

In general, the material response illustrated here and henceforth can be explained in terms of the competition on the minimization of the contributions governed by ℓ_1 and ℓ_2 in the defect energy. To shed light on this, it would be useful to analyze the plastic strain and plastic spin profiles, but this is skipped here for the sake of brevity. For this, the reader is referred to Bardella and Panteghini (2015), where the crucial relevance of accounting for the plastic spin is highlighted and it is also shown that all the three nonvanishing components of Nye's tensor (5.13) play an important role.

Figure 5.8 illustrates the well-known strengthening effect governed by the dissipative length scale, while Fig. 5.9 shows the very relevant influence of the parameter χ governing the dissipation due to the plastic spin. Of course, increasing χ makes the wire stronger. Let us remark that the case $\chi = 0$ does not correspond to vanishing plastic spin, as, in this case, the plastic spin is free to assume the most suitable form that can minimize the defect energy. In other words, the assumption $\chi = 0$ does not lead to the irrotational plastic flow condition adopted by Gurtin and Anand (2005).

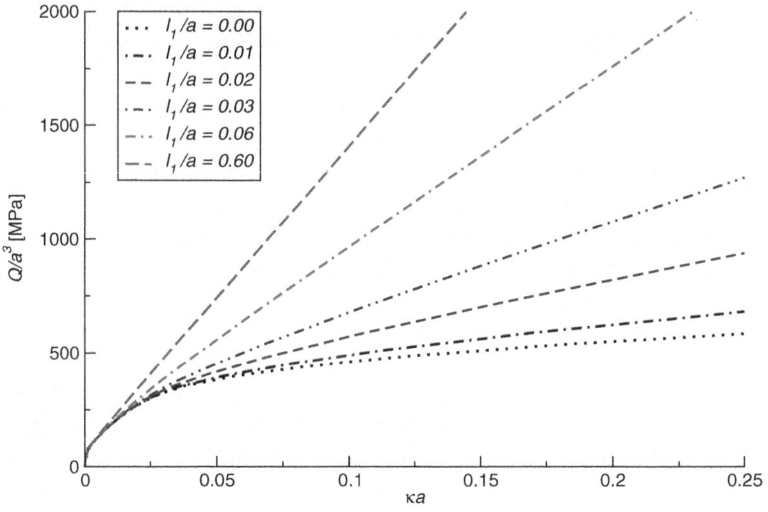

Fig. 5.6 Normalized torque versus applied twist: effect of the energetic length scale ℓ_1. Other parameters are $\ell_2/a = 0.3$, $L = 0$, $\chi = 2/3$. From Bardella and Panteghini (2015)

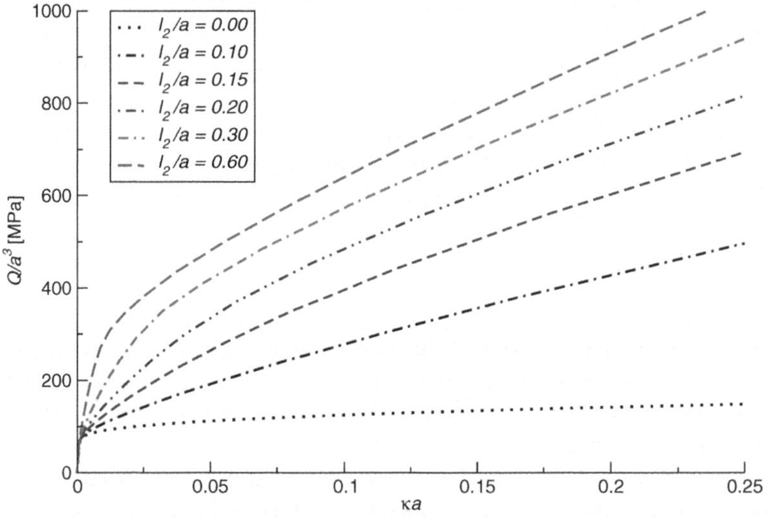

Fig. 5.7 Normalized torque versus applied twist: effect of the energetic length scale ℓ_2. Other parameters are $\ell_1/a = 0.02$, $L = 0$, $\chi = 2/3$. From Bardella and Panteghini (2015)

Now, we move to the case of non-proportional loading, still under the assumption of quadratic defect energy (5.74). In Figs. 5.10 and 5.11, we illustrate the behavior ensuing from the switch in higher-order boundary conditions, from microfree to

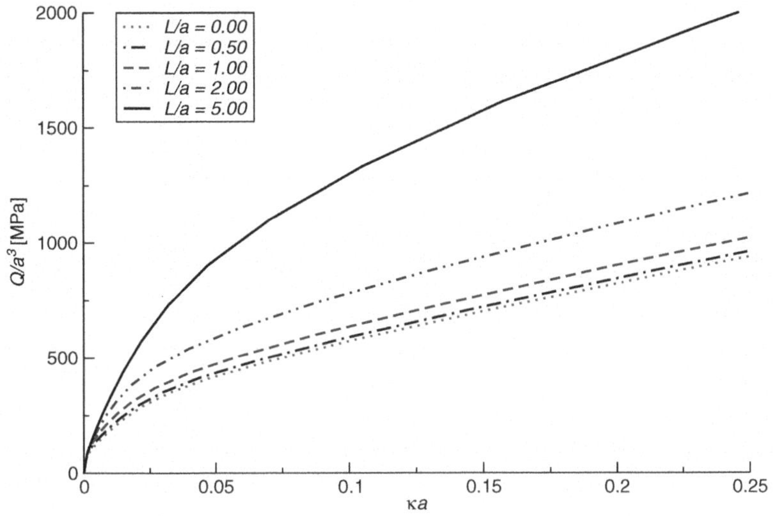

Fig. 5.8 Normalized torque versus applied twist: effect of the dissipative length scale L. Other parameters are $\ell_1/a = 0.02$, $\ell_2/a = 0.3$, $\chi = 2/3$. From Bardella and Panteghini (2015)

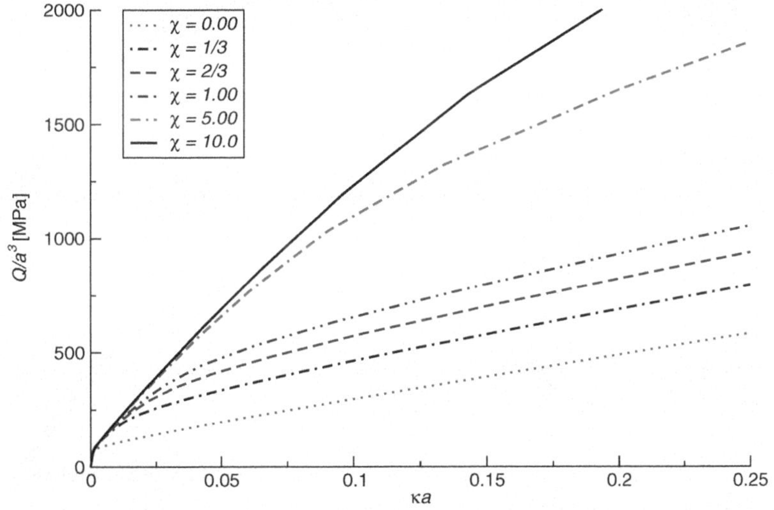

Fig. 5.9 Normalized torque versus applied twist: effect of the material parameter χ. Other parameters are $\ell_1/a = 0.02$, $\ell_2/a = 0.3$, $L = 0$. From Bardella and Panteghini (2015)

microhard, in the plastic regime, which is an important source of non-proportional loading and describes the sudden formation of a passivation layer (Fleck et al. 2014).

In this FE simulation, we have employed a finer mesh, constituted by 1000 elements of variable length: by referring to the smallest element at the boundary $r = a$,

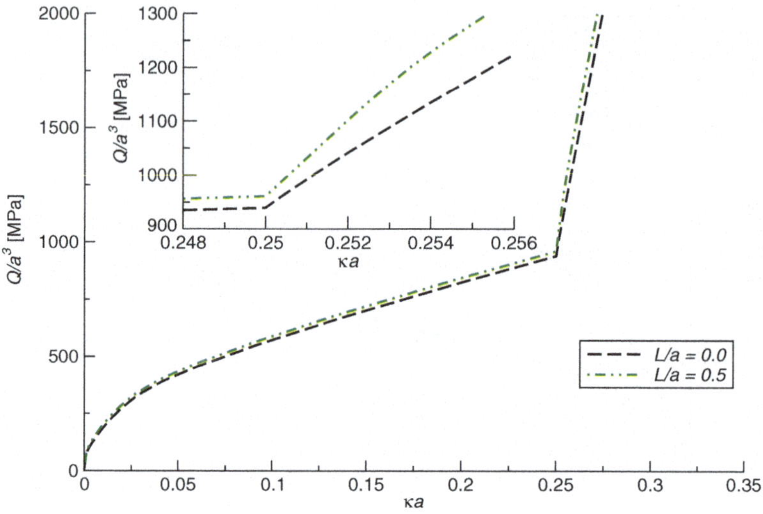

Fig. 5.10 Effect of the application of a passivation layer: normalized torque versus applied twist κa. From Bardella and Panteghini (2015)

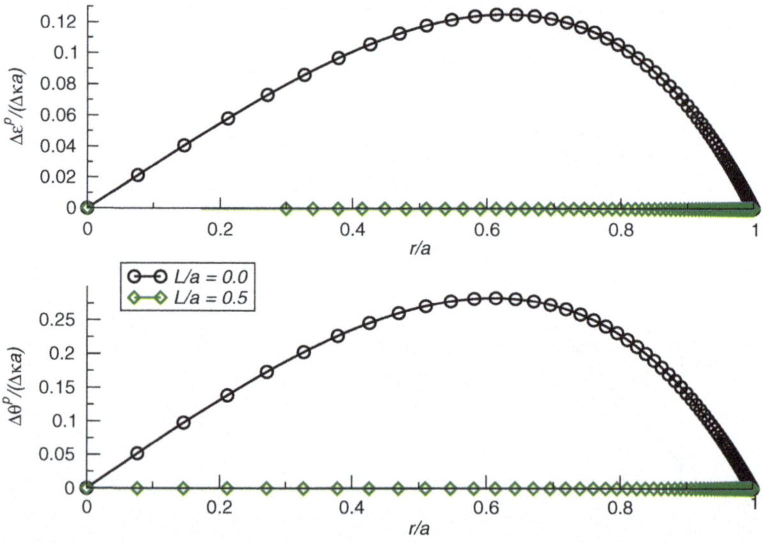

Fig. 5.11 Plastic response immediately after application of a passivation layer: normalized plastic strain and plastic spin increments along the radial coordinate. From Bardella and Panteghini (2015)

each element is 2% larger than that preceding it. The material parameters are chosen as follows: $\ell_1/a = 0.02$, $\ell_2/a = 0.3$, $\sigma_0 = 55$ MPa, $\dot{\varepsilon}_0 = 1 \cdot 10^{-4}$ s^{-1}, $\chi = 2/3$.

In particular, Figs. 5.10 and 5.11 illustrate the comparison between the cases with and without higher-order dissipation (the latter simply obtained by setting $L = 0$). In

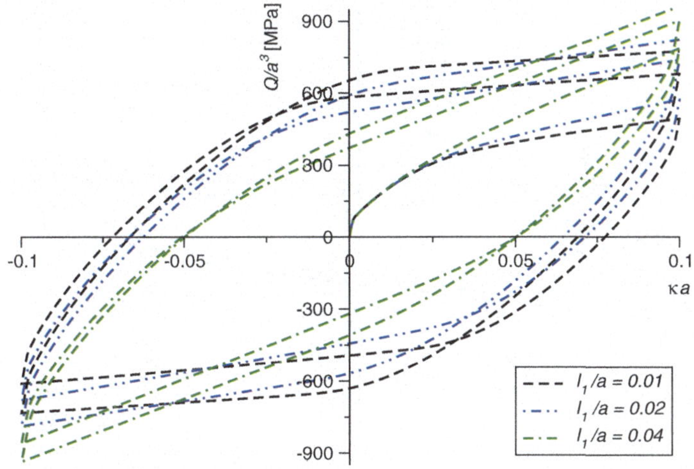

Fig. 5.12 Cyclic behavior: normalized torque versus applied twist as function of ℓ_1/a, with $\ell_2/a = 0.3$. From Bardella and Panteghini (2015)

the torque-twist curves of Fig. 5.10, little difference is observed in the two curves: both display a conspicuous, and apparently similar, stiffening when passivation occurs. However, Fig. 5.11 shows that the two responses are quite different, as, in the presence of higher-order dissipation when passivation occurs, the incremental response is linear elastic in the whole domain. This phenomenon is denoted as *elastic gap*, and it has been studied since the contribution of Fleck et al. (2014) (see Willis 2019 and references therein). As demonstrated by Fig. 5.11, the response in the absence of higher-order dissipation is free from elastic gap.

We further note that the elastic gap exists only in the case of rate-independent plasticity. Hence, this analysis also shows that the viscoplastic function (5.31) allows one to efficiently approach this limit.

Now, we consider the case of cyclic loading. Figure 5.12 demonstrates that the length scale ℓ_1, governing the influence of α_{ii}, controls the strain hardening slope within the cycle, whereas Fig. 5.13 demonstrates that the length scale ℓ_2, governing the influence of $|\alpha^{\mathrm{d}}|$, controls the cycle amplitude.

Finally, we consider the logarithmic defect energy (5.75) with the purpose of fitting the experimental results of Fleck et al. (1994) by neglecting higher-order dissipation (i.e., $L = 0$). Since the experimental results exhibit some strengthening, the logarithmic defect energy (5.75) is useful as it delivers a clear increase of defect stress for a relatively small Nye's tensor increment at small Nye's tensor value. This peculiarity, which characterizes the so-called *less-than-quadratic* defect energies, allows one to model a material exhibiting *energetic strengthening*, which actually consists of a very large size-dependent strain hardening at incipient plastic deformation.

The fitting of Fig. 5.14 is obtained with the following material parameters: $\sigma_0 = 68$ MPa, $\dot{\varepsilon}_0 = 1 \cdot 10^{-5}\ \mathrm{s}^{-1}$, $\ell_1 = 2.1 \cdot 10^{-4}\ \mu\mathrm{m}$, $\ell_2 = 2.1 \cdot 10^{-3}\ \mu\mathrm{m}$, $\alpha_1 \approx 2.8571 \cdot$

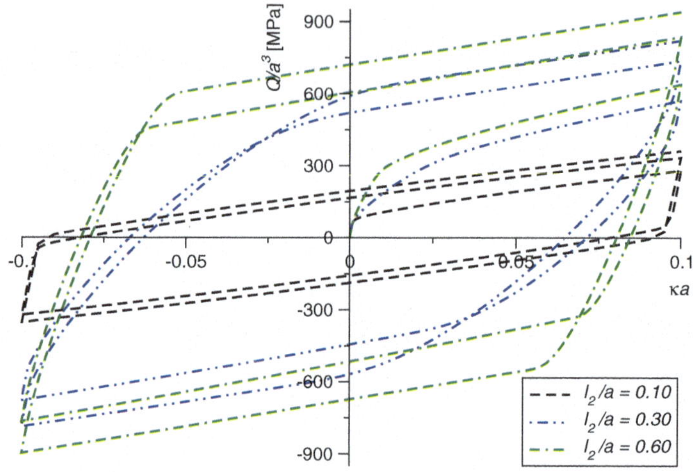

Fig. 5.13 Cyclic behavior: normalized torque versus applied twist as function of ℓ_2/a, with $\ell_1/a = 0.02$. From Bardella and Panteghini (2015)

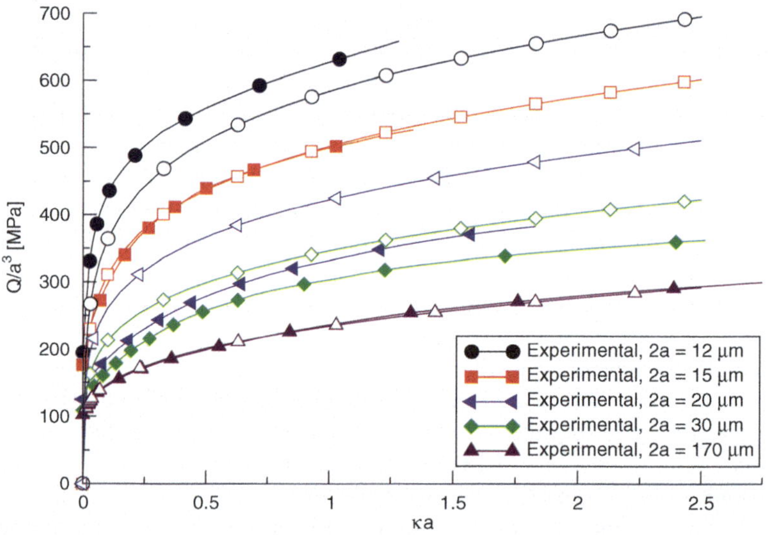

Fig. 5.14 Fitting the experimental results of Fleck et al. (1994) by adopting the logarithmic defect energy (5.75). Void symbols represent FE results. From Bardella and Panteghini (2015)

$10^{-4}, \alpha_2 \approx 5.7143 \cdot 10^{-3}, L = 0, \chi = 2/3$. The model and experimental results agree well, even without resorting to any optimized identification procedure. However, it is important to note that, as illustrated in Fig. 5.15 for the wire with diameter $2a = 12$ μm, less-than-quadratic defect energies may predict a cyclic behavior affected by a change of concavity which is rarely observed in polycrystalline metals. In particular,

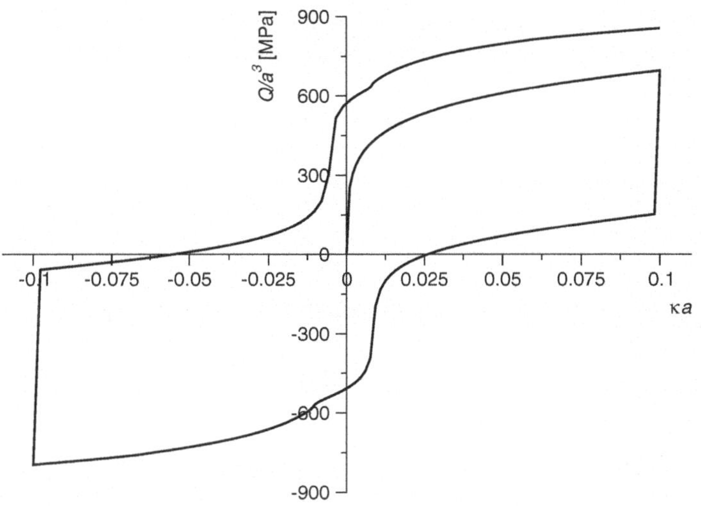

Fig. 5.15 Cyclic behavior: normalized torque versus applied twist by adopting the logarithmic defect energy (5.75). From Bardella and Panteghini (2015)

it is not displayed by the results of Liu et al. (2013) on the torsion of thin metal wires. We note that in the context of *crystal* plasticity, by referring to single-slip plasticity, Wulfinghoff et al. (2015) have explained this behavior by resorting to stage III kinematic hardening (Asaro 1975). More in this regard can be found in the recent investigation of El-Naaman et al. (2019).

Finally, we mention the very recent constitutive proposal of Panteghini et al. (2019), who put forward a thermodynamically consistent potential for strain gradient plasticity, which includes both energetic and dissipative higher-order contributions and allows the modeling of both strengthening and increase in strain hardening with diminishing size without being affected by "elastic gaps" nor cyclic behavior that may be questionable in polycrystal plasticity.

Strain Gradient Plasticity

We now specialize in the development of a *strain* gradient plasticity theory in a strict sense, for which plastic strain gradients are associated with non-conventional stresses. Although energetic gradient terms may be added to the formulation, by assuming, in the free energy density (5.32), a defect energy function of the current value of the plastic strain gradient, $\mathcal{D}(\nabla \varepsilon^p)$, as for instance in Gudmundson (2004), and Fleck and Willis (2009b), here we focus on the purely dissipative formulation, as it can be readily extended to finite strains (Niordson and Tvergaard 2019).

By assuming symmetry in the first two indices for all the static tensor fields in the general form of the internal and external virtual works (5.58)–(5.59)

$$\varsigma_{ij} = \varsigma_{ji}\,, \quad S_{ijk} = S_{jik}\,, \quad t_{ij} = t_{ji}$$

and by adopting the notation

$$q_{ij} = \varsigma_{ij} = \varsigma_{ji}\,, \quad m_{ijk} = S_{ijk} = S_{jik}$$

the *generalized* PVW (5.20) for any subregion $\Pi \subset \Omega$ can be stated as

$$\int_{\Pi} \left[\sigma_{ij}\delta\varepsilon_{ij} + (q_{ij} - \sigma^{\mathrm{d}}_{ij})\delta\varepsilon^p_{ij} + m_{ijk}\delta\varepsilon^p_{ij,k}\right]\mathrm{d}V = \int_{\partial\Pi} \left[T_i\delta u_i + t_{ij}\delta\varepsilon^p_{ij}\right]\mathrm{d}A \quad (5.80)$$

Here, q_{ij} is the non-conventional stress work conjugate to the plastic strain rate $\dot{\varepsilon}^p_{ij}$ and m_{ijk} is the higher-order stress which is work conjugate to the gradient of the plastic strain rate, $\dot{\varepsilon}^p_{ij,k}$. Surface tractions include the conventional traction vector T_i, conjugate to the displacement rate, and the higher-order traction t_{ij}, conjugate to the plastic strain rate. Application of the divergence theorem, while enforcing the PVW (5.20) to hold for all kinematically admissible variations in total strain and plastic strain, leads to the conventional equilibrium conditions (5.25) and (5.26) in addition to the higher-order conditions in the entire body

$$q_{ij} - m_{ijk,k} - \sigma^{\mathrm{d}}_{ij} = 0 \quad \text{in } \Omega$$

and over the surface where the higher-order tractions are known

$$t_{ij} = m_{ijk}n_k \quad \text{on } S^{\mathrm{dis}}_T$$

Dissipation is assumed to be due to a gradient enhanced effective plastic strain measure defined incrementally by

$$\dot{E}^p = \sqrt{\frac{2}{3}\dot{\varepsilon}^p_{ij}\dot{\varepsilon}^p_{ij} + (L_D)^2\dot{\varepsilon}^p_{ij,k}\dot{\varepsilon}^p_{ij,k}} \quad (5.81)$$

where L_D is a dissipative length parameter setting the scale for gradient effects. We remark that here L_D plays the same role of $L_\perp$, $L_\odot$, and L_M in (5.40) for the crystal case and of L in Eq. (5.69) for distortion gradient plasticity. Note that, differently from definition (5.69), in Eq. (5.81) the higher-order contribution is not multiplied by the 2/3 factor.

In this purely dissipative SGP, for a vanishing length parameter L_D in definition (5.81) we recover the conventional effective plastic strain rate. The work conjugate effective stress quantity is given by

$$\Sigma = \sqrt{\frac{3}{2} q_{ij} q_{ij} + \frac{1}{(L_D)^2} m_{ijk} m_{ijk}}$$

We may straightforwardly obtain numerical solutions in a finite element setting, from the *generalized* PVW (5.20), by solving for increments of displacements and increments of plastic strain *rate* (Fredriksson and Gudmundson 2005; Niordson and Legarth 2010). Alternatively, FE solutions can be formulated by using as degrees of freedom increments of the displacements and the plastic strain rate itself, based on the minimum principles formulated by Fleck and Willis (2009b).

Setting $\delta u_i = 0$ in (5.80) leads to the variational principle

$$\int_{\Pi} \left(q_{ij} \delta \varepsilon^p_{ij} + m_{ijk} \delta \varepsilon^p_{ij,k} \right) \mathrm{d}V = \int_{\Pi} \sigma^{\mathrm{d}}_{ij} \delta \varepsilon^p_{ij} \mathrm{d}V + \int_{\partial \Pi} t_{ij} \delta \varepsilon^p_{ij} \mathrm{d}S \tag{5.82}$$

which may form the basis for finding incremental plastic strain fields. This variational principle is consistent with the viscoplastic minimum principle proposed by Fleck and Willis (2009b)

$$\inf_{\dot{\varepsilon}^{p*}_{ij}} \left\{ \int_{\Pi} \left(\mathcal{V}(E^p, \dot{E}^p) - \sigma^{\mathrm{d}}_{ij} \dot{\varepsilon}^{p*}_{ij} \right) \mathrm{d}V - \int_{\partial \Pi} t_{ij} \dot{\varepsilon}^{p*}_{ij} \mathrm{d}A \right\}$$

Here, the viscoplastic potential $\mathcal{V}(E^p, \dot{E}^p)$ is given by Eq. (5.72).

Setting $\delta \varepsilon^p_{ij} = 0$ in (5.80) leads to the conventional PVW, given by

$$\int_{\Pi} \sigma_{ij} \delta \varepsilon_{ij} \, \mathrm{d}V = \int_{\partial \Pi} T_i \delta u_i \, \mathrm{d}S \tag{5.83}$$

In a finite strain context, the variational statements (5.82) and (5.83) must be fulfilled in the deformed configuration, and Π is thus a subregion of that configuration (Niordson and Tvergaard 2019). In the present setting, the dissipative stresses are given in terms of the plastic strain rate field according to

$$q_{ij} = \frac{2}{3} \frac{\Sigma}{\dot{E}^p} \dot{\varepsilon}^p_{ij} \,, \quad m_{ijk} = (L_D)^2 \frac{\Sigma}{\dot{E}^p} \dot{\varepsilon}^p_{ij,k} \tag{5.84}$$

and a higher-order equilibrium solution is obtained directly by finding a plastic strain rate field that satisfies (5.82) in the deformed state, given the stress deviator field and appropriate boundary conditions. We remark the analogy in the dissipative constitutive laws within SGCP, distortion gradient plasticity, and the present phenomenological SGP, given by Eqs. (5.41), (5.70), and (5.84), respectively. This structure of the constitutive equations allows one to easily satisfy the second law of thermodynamics.

By following McMeeking and Rice (1975), we adopt an updated Lagrangian framework in which the conventional PVW is transformed to an updated reference configuration, Π_0, using the first Piola–Kirchhoff stress, P_{ij}, and the nominal traction, T_i^{N}, according to

$$\int_{\Pi_0} P_{ij}\delta u_{j,i}\mathrm{d}V = \int_{\partial\Pi_0} T_i^{\mathrm{N}}\delta u_i \mathrm{d}A$$

By time differentiation, we obtain the *incremental* PVW

$$\int_{\Pi_0} \dot{P}_{ij}\delta u_{j,i}\mathrm{d}V = \int_{\delta\Pi_0} \dot{T}_i^{\mathrm{N}}\delta u_i \mathrm{d}A$$

where the increment of the first Piola–Kirchhoff stress can be related to the Jaumann rate of the Kirchhoff stress τ_{ij} by

$$\dot{P}_{ij} = \overset{\triangledown}{\tau}_{ij} - \sigma_{kj}\dot{\varepsilon}_{ki} - \sigma_{ik}\dot{\epsilon}_{kj} + \sigma_{ik}\dot{u}_{j,k}$$

which results in the virtual work statement

$$\int_{\Pi_0} \left(\overset{\triangledown}{\tau}_{ij}\delta\varepsilon_{ij} - \sigma_{ij}\left(2\dot{\varepsilon}_{ik}\delta\varepsilon_{kj} - \dot{u}_{k,j}\delta u_{k,i}\right)\right)\mathrm{d}V = \int_{\partial\Pi_0} \dot{T}_i^{\mathrm{N}}\delta u_i \mathrm{d}A \tag{5.85}$$

As in McMeeking and Rice (1975), we assume a hypo-elastic constitutive law defining the relationship between the Jaumann rate of the Kirchhoff stress and the elastic strain increment according to

$$\overset{\triangledown}{\tau}_{ij} = \dot{\tau}_{ij} - \dot{\vartheta}_{ik}\tau_{kj} - \tau_{ik}\dot{\vartheta}_{kj} = L_{ijkl}(\dot{\varepsilon}_{kl} - \dot{\varepsilon}^p_{kl}) \tag{5.86}$$

where $\vartheta_{ij} = (u_{i,j} - u_{j,i})/2$ and L_{ijkl} is assumed to be isotropic (see Eq. (5.28)). Substitution of the constitutive relation (5.86) into the virtual statement (5.85) leads to the incremental variational relation

$$\int_{\Pi_0} \left[L_{ijkl}\dot{\varepsilon}_{kl}\delta\varepsilon_{ij} - \sigma_{ij}\left(2\dot{\varepsilon}_{ik}\delta\varepsilon_{kj} - \dot{u}_{k,j}\delta u_{k,i}\right)\right]\mathrm{d}V$$
$$= \int_{\partial\Pi_0} \left(L_{ijkl}\dot{\varepsilon}^p_{kl}\delta\varepsilon_{ij} + \dot{T}_i^{\mathrm{N}}\delta u_i\right)\mathrm{d}A \tag{5.87}$$

Given a known plastic strain rate field from (5.82), the velocity field may be determined from the incremental virtual work statement (5.87).

Finite Element Implementation

FE solutions based on the minimum principles of Fleck and Willis (2009b) may be found by adopting the numerical methods developed in Niordson and Hutchinson (2011), Danas et al. (2012), and Nielsen and Niordson (2014). Finite strain formulations are developed in Niordson and Tvergaard (2018) and Nielsen and Niordson (2019).

While the following FE formulation is restricted to viscoplasticity, time-independent solutions may be obtained from formulations making use of image analysis to determine disjunct plastic zones (Nielsen and Niordson 2014, 2019).

We interpolate displacement increments and strain rates by using FE shape functions as follows

$$\Delta u_i = \sum_{n=1}^{N_{II}} N_i^{(n)} \Delta U^{(n)} \quad \text{and} \quad \dot{\varepsilon}_{ij}^p = \sum_{n=1}^{N_I} M_{ij}^{(n)} \dot{\varepsilon}^{(n)}$$

where $N_i^{(n)}$ and $M_{ij}^{(n)}$ are shape functions and $\Delta U^{(n)}$ and $\dot{\varepsilon}^{(n)}$ are nodal degrees of freedom.

By introducing this discretization in (5.82) along with the constitutive laws (5.84), we obtain

$$\int_\Pi \frac{\Sigma}{\dot{E}_k^p} \left(\frac{2}{3} M_{ij}^{(n)} M_{ij}^{(m)} + L_D^2 M_{ij,k}^{(n)} M_{ij,k}^{(m)} \right) \mathrm{d}V \cdot \dot{\varepsilon}_{k+1}^{p(m)} = \int_{\partial\Pi} \sigma_{ij}^{\mathrm{d}} M_{ij}^{(n)} \mathrm{d}V + \int_{S_e} t_{ij} M_{ij}^{(n)} \mathrm{d}A \tag{5.88}$$

We solve Eq. (5.88) iteratively, $k = 1, 2, \ldots$, for the nodal values of the plastic strain rate. Based on a known plastic strain rate field, we find the displacement increments from the discretized version of (5.87), with plastic contributions on the right-hand side:

$$\int_\Pi \left[L_{ijkl} B_{ij}^{(n)} B_{kl}^{(m)} + \sigma_{ij} \left(N_{k,j}^{(m)} N_{k,i}^{(n)} - 2 B_{ik}^{(m)} B_{kj}^{(n)} \right) \right] \mathrm{d}V \cdot \Delta U^{(m)} = \Delta t \left(\int_\Pi L_{ijkl} B_{ij}^{(n)} \dot{\varepsilon}_{kl}^p \mathrm{d}V + \int_{\partial\Pi} \dot{T}_i N_i^{(n)} \mathrm{d}A \right) \tag{5.89}$$

Here, $B_{ij}^{(n)} = (N_{i,j}^{(n)} + N_{j,i}^{(n)})/2$ and Δt is the time increment.

The flow of the solution method is as follows:

1. We find the incremental elastic solution from Eq. (5.89).
2. We incrementally update the Kirchhoff stresses according to Eq. (5.86), and then we calculate the Cauchy stresses by the relation $\sigma_{ij} = u_{k,k}\tau_{ij}$.
3. We update the nodal coordinates.
4. We obtain the nodal values for the plastic strain rate field from Eq. (5.88).

The above is repeated, thus time integrating through the loading history.

References

Acharya, A., & Bassani, J. L. (2000). Lattice incompatibility and a gradient theory of crystal plasticity. *Journal of the Mechanics and Physics of Solids*, *48*, 1565–1595.

Anand, L., Gurtin, M. E., Lele, S. P., & Gething, C. (2005). A one-dimensional theory of strain-gradient plasticity: Formulation, analysis, numerical results. *Journal of the Mechanics and Physics of Solids*, *53*, 1789–1826.

Arsenlis, A., & Parks, D. M. (1999). Crystallographic aspects of geometrically-necessary and statistically-stored dislocation density. *Acta Materialia*, *47*(5), 1597–1611.

Asaro, R. J. (1975). Elastic-plastic memory and kinematic-type hardening. *Acta Metallurgica*, *23*, 1255–1265.

Ashby, M. F. (1970). The deformation of plastically non-homogeneous materials. *Philosophical Magazine*, *21*, 399–424.

Bardella, L. (2006). A deformation theory of strain gradient crystal plasticity that accounts for geometrically necessary dislocations. *Journal of the Mechanics and Physics of Solids*, *54*, 128–160.

Bardella, L. (2007). Some remarks on the strain gradient crystal plasticity modelling, with particular reference to the material length scales involved. *International Journal of Plasticity*, *23*, 296–322.

Bardella L. (2020). *Strain Gradient Plasticity*. In H. Altenbach & A. Öchsner (Eds), *Encyclopedia of Continuum Mechanics*. Springer-Verlag, Berlin Heidelberg, pp. 2330–2341. https://doi.org/10.1007/978-3-662-55771-6_110

Bardella, L., & Giacomini, A. (2008). Influence of material parameters and crystallography on the size effects describable by means of strain gradient plasticity. *Journal of the Mechanics and Physics of Solids*, *56*(9), 2906–2934.

Bardella, L., & Panteghini, A. (2015). Modelling the torsion of thin metal wires by distortion gradient plasticity. *Journal of the Mechanics and Physics of Solids*, *78*, 467–492.

Bardella, L., Segurado, J., Panteghini, A., & Llorca, J. (2013). Latent hardening size effect in small-scale plasticity. *Modelling and Simulation in Materials Science and Engineering*, *21*(5), 055009.

Bishop, J. F. W., & Hill, R. (1951). A theory of plastic distortion of a polycrystalline aggregate under combined stress. *Philosophical Magazine*, *42*, 414–427.

Bittencourt, E., Needleman, A., Gurtin, M. E., & van der Giessen, E. (2003). A comparison of nonlocal continuum and discrete dislocation plasticity predictions. *Journal of the Mechanics and Physics of Solids*, *51*, 281–310.

Borg, U., & Fleck, N. A. (2007). Strain gradient effects in surface roughening. *Modelling and Simulation in Materials Science and Engineering*, *15*, 1–12.

Burgers, J. M. (1939). Some considerations of the field of stress connected with dislocations in a regular crystal lattice. *Koninklijke Nederlandse Akademie Van Wetenschappen*, *42*, 293–325 (Part 1), 378–399 (Part 2).

Chiricotto, M., Giacometti, L., & Tomassetti, G. (2012). Torsion in strain-gradient plasticity: Energetic scale effects. *SIAM Journal on Applied Mathematics*, *72*(4), 1169–1191.

Chiricotto, M., Giacometti, L., & Tomassetti, G. (2016). Dissipative scale effects in strain-gradient plasticity: The case of simple shear. *SIAM Journal on Applied Mathematics*, *76*(2), 688–704.

Conti, S., & Ortiz, M. (2005). Dislocation microstructures and the effective behavior of single crystals. *Archive for Rational Mechanics and Analysis*, *176*, 103–147.

Danas, K., Deshpande, V. S., & Fleck, N. A. (2012). Size effects in the conical indentation of an elasto-plastic solid. *Journal of the Mechanics and Physics of Solids*, *60*(9), 1605–1625.

Del Piero, G. (2009). On the method of virtual power in continuum mechanics. *Journal of Mechanics of Materials and Structures*, *4*(2), 281–292.

El-Naaman, S. A., Nielsen, K. L., & Niordson, C. F. (2019). An investigation of back stress formulations under cyclic loading. *Mechanics of Materials*, *130*, 76–87.

Ertürk, I., van Dommelen, J. A. W., & Geers, M. G. D. (2009). Energetic dislocation interactions and thermodynamical aspects of strain gradient crystal plasticity theories. *Journal of the Mechanics and Physics of Solids*, *57*(11), 1801–1814.

Evans, A. G., & Hutchinson, J. W. (2009). A critical assessment of theories of strain gradient plasticity. *Acta Materialia*, *57*(5), 1675–1688.

Evers, L. P., Brekelmans, W. A. M., & Geers, M. G. D. (2004). Non-local crystal plasticity model with intrinsic SSD and GND effects. *Journal of the Mechanics and Physics of Solids*, *52*(10), 2379–2401.

Fleck, N. A., & Hutchinson, J. W. (1997). Strain gradient plasticity. *Advances in Applied Mechanics*, *33*, 295–361.

Fleck, N. A., & Hutchinson, J. W. (2001). A reformulation of strain gradient plasticity. *Journal of the Mechanics and Physics of Solids*, *49*(10), 2245–2271.

Fleck, N. A., & Willis, J. R. (2009a). A mathematical basis for strain-gradient plasticity theory. Part I: Scalar plastic multiplier. *Journal of the Mechanics and Physics of Solids*, *57*, 161–177.

Fleck, N. A., & Willis, J. R. (2009b). A mathematical basis for strain-gradient plasticity theory. Part II: Tensorial plastic multiplier. *Journal of the Mechanics and Physics of Solids*, *57*, 1045–1057.

Fleck, N. A., Muller, G. M., Ashby, M. F., & Hutchinson, J. W. (1994). Strain gradient plasticity: Theory and experiments. *Acta Metallurgica et Materialia*, *42*, 475–487.

Fleck, N. A., Hutchinson, J. W., & Willis, J. R. (2014). Strain gradient plasticity under non-proportional loading. *Proceedings of the Royal Society of London A*, *470*, 20140267.

Fleck, N. A., Hutchinson, J. W., & Willis, J. R. (2015). Guidelines for constructing strain gradient plasticity theories. *Journal of Applied Mechanics, Transactions ASME*, *82*, 071002, 1–10.

Forest, S., & Guéninchault, N. (2013). Inspection of free energy functions in gradient crystal plasticity. *Acta Mechanica Sinica*, *29*(6), 763–772.

Forest, S., & Sievert, R. (2003). Elastoviscoplastic constitutive frameworks for generalized continua. *Acta Mechanica*, *160*, 71–111.

Fredriksson, P., & Gudmundson, P. (2005). Size-dependent yield strength of thin films. *International Journal of Plasticity*, *21*, 1834–1854.

Gambin, W. (1992). Refined analysis of elastic-plastic crystals. *International Journal of Solids and Structures*, *29*, 2013–2021.

Gao, H., Huang, Y., Nix, W. D., & Hutchinson, J. W. (1999). Mechanism-based strain gradient plasticity — I. Theory. *Journal of the Mechanics and Physics of Solids*, *47*, 1239–1263.

Garroni, A., Leoni, G., & Ponsiglione, M. (2010). Gradient theory for plasticity via homogenization of discrete dislocations. *Journal of the European Mathematical Society*, *12*(5), 1231–1266.

Groma, I., Csikor, F. F., & Zaiser, M. (2003). Spatial correlations and higher-order gradient terms in a continuum description of dislocation dynamics. *Acta Materialia*, *51*, 1271–1281.

Groma, I., Györgyi, G., & Kocsis, B. (2007). Dynamics of coarse grained dislocation densities from an effective free energy. *Philosophical Magazine*, *87*(8–9), 1185–1199.

Gudmundson, P. (2004). A unified treatment of strain gradient plasticity. *Journal of the Mechanics and Physics of Solids*, *52*, 1379–1406.

Gurtin, M. E. (2002). A gradient theory of single-crystal viscoplasticity that accounts for geometrically necessary dislocations. *Journal of the Mechanics and Physics of Solids*, *50*, 5–32.

Gurtin, M. E. (2004). A gradient theory of small-deformation isotropic plasticity that accounts for the Burgers vector and for dissipation due to plastic spin. *Journal of the Mechanics and Physics of Solids*, *52*, 2545–2568.

Gurtin, M. E., & Anand, L. (2005). A gradient theory of strain-gradient plasticity for isotropic, plastically irrotational materials. Part I: Small deformations. *Journal of the Mechanics and Physics of Solids*, *53*, 1624–1649.

Gurtin, M. E., & Anand, L. (2007). A gradient theory for single-crystal plasticity. *Modelling and Simulation in Materials Science and Engineering*, *15*, 263–270.

Gurtin, M. E., & Anand, L. (2009). Thermodynamics applied to gradient theories involving the accumulated plastic strain: The theories of Aifantis and Fleck & Hutchinson and their generalization. *Journal of the Mechanics and Physics of Solids*, *57*, 405–421.

Gurtin, M. E., & Needleman, A. (2005). Boundary conditions in small-deformation, single-crystal plasticity that account for the Burgers vector. *Journal of the Mechanics and Physics of Solids*, *53*, 1–31.

Gurtin, M. E., & Ohno, N. (2011). A gradient theory of small-deformation, single crystal plasticity that accounts for GND-induced interactions between slip systems. *Journal of the Mechanics and Physics of Solids*, *59*, 320–343.

Gurtin, M. E., & Reddy, B. D. (2014). Gradient single-crystal plasticity within a Mises-Hill framework based on a new formulation of self- and latent-hardening. *Journal of the Mechanics and Physics of Solids*, *68*, 134–160.

Gurtin, M. E., Anand, L., & Lele, S. P. (2007). Gradient single-crystal plasticity with free energy dependent on dislocation densities. *Journal of the Mechanics and Physics of Solids*, *55*, 1853–1878.

Hall, E. O. (1951). The deformation and ageing of mild steel: III discussion of results. *Proceedings of the Physical Society B*, *64*, 747–753.

Hull, D., & Bacon, D. J. (2001). *Introduction to dislocations* (4th ed.). Oxford: Butterworth-Heinemann.

Hutchinson, J. W. (2000). Plasticity at the micron scale. *International Journal of Solids and Structures*, *37*, 225–238.

Kraft, O., Hommel, M., & Arzt, E.: X-ray diffraction as a tool to study the mechanical behaviour of thin films. *Materials Science and Engineering: A*, *288*, 209–216.

Kröner, E. (1962). Dislocations and continuum mechanics. *Applied Mechanics Reviews*, *15*, 599–606.

Kuroda, M., & Tvergaard, V. (2008). On the formulations of higher-order strain gradient crystal plasticity models. *Journal of the Mechanics and Physics of Solids*, *56*, 1591–1608.

Lanczos, C. (1970). *The variational principles of mechanics* (4th ed.). Toronto: University of Toronto Press.

Liu, D., He, Y., Dunstan, D. J., Zhang, B., Gan, Z., Hu, P., et al. (2013). Anomalous plasticity in cyclic torsion of micron scale metallic wires. *Physical Review Letters*, *110*, 244301.

Ma, Q., & Clarke, D. R. (1995). Size dependent hardness in silver single crystals. *Journal of Materials Research*, *10*, 853–863.

Malvern, L. E. (1969). *Introduction to the mechanics of a continuous medium*. New Jersey: Prentice-Hall.

Martínez-Pañeda, E., Niordson, C. F., & Bardella, L. (2016). A finite element framework for distortion gradient plasticity with applications to bending of thin foils. *International Journal of Solids and Structures*, *96*, 288–299.

Martínez-Pañeda, E., Deshpande, V. S., Niordson, C. F., & Fleck, N. A. (2019). The role of plastic strain gradients in the crack growth resistance of metals. *Journal of the Mechanics and Physics of Solids*, *126*, 136–150.

McMeeking, R. M., & Rice, J. R. (1975). Finite-element formulations for problems of large elastic-plastic deformation. *International Journal of Solids and Structures*, *11*, 601–616.

Moreau, P., Raulic, M., P'ng, M. Y., Gannaway, G., Anderson, P., Gillin, W. P., et al. (2005). Measurement of the size effect in the yield strength of nickel foils. *Philosophical Magazine Letters*, *85*(7), 339–343.

Needleman, A. (1988). Material rate dependence and mesh sensitivity in localization problems. *Computer Methods in Applied Mechanics and Engineering*, *67*, 69–85.

Nicola, L., van der Giessen, E., & Gurtin, M. E. (2005). Effect of defect energy on strain-gradient predictions of confined single-crystal plasticity. *Journal of the Mechanics and Physics of Solids*, *53*, 1280–1294.

Nielsen, K. L., & Niordson, C. F. (2014). A numerical basis for strain-gradient plasticity theory: Rate-independent and rate-dependent formulations. *Journal of the Mechanics and Physics of Solids*, *63*, 113–127.

Nielsen, K. L., & Niordson, C. F. (2019). Print share a finite strain FE-implementation of the Fleck-Willis gradient theory: Rate-independent versus visco-plastic formulation. *European Journal of Mechanics/A Solids*, *75*, 389–398.

Niordson, C. F., & Hutchinson, J. W. (2011). Basic gradient plasticity theories with application to constrained film deformation. *Journal of Mechanics of Materials and Structures*, *6*(1–4), 395–416.

Niordson, C. F., & Legarth, B. N. (2010). Strain gradient effects on cyclic plasticity. *Journal of the Mechanics and Physics of Solids*, *58*, 542–557.

Niordson, C. F., & Tvergaard, V. (2018). A homogenized model for size-effects in porous metals. *Journal of the Mechanics and Physics of Solids*, *123*, 222–233.

Niordson, C. F., & Tvergaard, V. (2019). A homogenized model for size-effects in porous metals. *Journal of the Mechanics and Physics of Solids*, *123*, 222–233.

Nye, J. F. (1953). Some geometrical relations in dislocated crystals. *Acta Metallurgica*, *1*, 153–162.

Ohno, N., & Okumura, D. (2007). Higher-order stress and grain size effects due to self-energy of geometrically necessary dislocations. *Journal of the Mechanics and Physics of Solids*, *55*, 1879–1898.

Ortiz, M., & Popov, E. P. (1985). Accuracy and stability of integration algorithms for elastoplastic constitutive relations. *International Journal for Numerical Methods in Engineering*, *21*(9), 1561–1576.

Panteghini, A., & Bardella, L. (2016). On the finite element implementation of higher-order gradient plasticity, with focus on theories based on plastic distortion incompatibility. *Computer Methods in Applied Mechanics and Engineering*, *310*, 840–865.

Panteghini, A., & Bardella, L. (2018). On the role of higher-order conditions in distortion gradient plasticity. *Journal of the Mechanics and Physics of Solids*, *118*, 293–321. ISSN 0022-5096.

Panteghini, A., Bardella, L., & Niordson, C. F. (2019). A potential for higher-order phenomenological strain gradient plasticity to predict reliable response under non-proportional loading. *Proceedings of the Royal Society of London A*, *475*(2229), 20190258.

Peirce, D., Asaro, R. J., & Needleman, A. (1983). Material rate dependence and localized deformation in crystalline solids. *Acta Metallurgica*, *31*(12), 1951–1976.

Petch, N. J. (1953). The cleavage strength of polycrystals. *Journal of the Iron and Steel Institute*, *174*, 25–28.

Poh, L. H., & Peerlings, R. H. J. (2016). The plastic rotation effect in an isotropic gradient plasticity model for applications at the meso scale. *International Journal of Solids and Structures*, *78–79*, 57–69.

Polizzotto, C. (2009). A link between the residual-based gradient plasticity theory and the analogous theories based on the virtual work principle. *International Journal of Plasticity*, *25*, 2169–2180.

Polizzotto, C., Borino, G., & Fuschi, P. (1998). A thermodynamically consistent formulation of nonlocal and gradient plasticity. *Mechanics Research Communications*, *25*(1), 75–82.

Stelmashenko, N. A., Walls, M. G., Brown, L. M., & Milman, Y. V. (1993). Microindentations on W and Mo oriented single crystals: An STM study. *Acta Metallurgica et Materialia*, *41*, 2855–2865.

Stölken, J. S., & Evans, A. G. (1998). A microbend test method for measuring the plasticity length scale. *Acta Materialia*, *46*, 5109–5115.

Svendsen, B., & Bargmann, S. (2010). On the continuum thermodynamic rate variational formulation of models for extended crystal plasticity at large deformation. *Journal of the Mechanics and Physics of Solids*, *58*, 1253–1271.

Willis, J. R. (2019). Some forms and properties of models of strain-gradient plasticity. *Journal of the Mechanics and Physics of Solids*, *123*, 348–356.

Wulfinghoff, S., Forest, S., & Böhlke, T. (2015). Strain gradient plasticity modelling of the cyclic behaviour of laminate microstructures. *Journal of the Mechanics and Physics of Solids*, *79*, 1–20.

Xiang, Y., & Vlassak, J. J. (2006). Bauschinger and size effects in thin-film plasticity. *Acta Materialia*, *54*, 5449–5460.

Chapter 6
Finite Gradient Elasticity and Plasticity

Albrecht Bertram

Abstract Within the range of finite deformations, material laws have to fulfil certain invariance requirements. If we assume invariance under rigid body modifications, it is generally recommended to use material variables because of their invariance property, i.e. those which are defined in the reference placement. For the higher-order variables, in our case triadics, we introduce the configuration tensor and a generalized version of the Piola–Kirchhoff stress tensor. With these variables, we can produce reduced forms. In the case of elasticity, such reduced forms depend on the reference placement. Its transformation behaviour is the key to introduce isomorphy and symmetry conditions. As an example, we consider a gradient version of the St. Venant–Kirchhoff law, i.e. a linear isotropic second gradient elastic law. In the second part of this chapter, a format for a gradient elastoplasticity is given. The basic concept here is that of elastic ranges. This is a subset of the space of configurations within which the behaviour is elastic. Only under yielding, the elastic ranges evolve which is described by flow and hardening rules. A general rate-independent ansatz for these rules is given. A loading condition and a consistency condition can then be derived. This completes the general model of an elastoplastic second-order material.

Keywords Curvature · Elastic range · Flow rule · Hardening rule · St. Venant-Kirchhoff law · Yield criterion

Introduction

While gradient elasticity and plasticity became rather popular during the last years, large deformations were rather seldom implemented into such formats. This is surely not due to a lesser need for it. On the contrary, gradients become particularly important under large deformations. The real reason is probably that gradient theories become much more complicated under large plastic deformations. One has to take into account certain invariance requirements which are commonly neglected in linear

A. Bertram (✉)
Technical University Berlin, Otto-von-Guericke-University, Magdeburg, Germany
e-mail: albrecht.bertram@ovgu.de

A. Bertram and S. Forest (eds.), *Mechanics of Strain Gradient Materials*,
CISM International Centre for Mechanical Sciences 600,
https://doi.org/10.1007/978-3-030-43830-2_6

theories. So, even the choice of appropriate kinematical and dynamical variables is not trivial for finite theories.

Rare examples for the consideration of finite gradient elasticity are given by Cross (1973), Murdoch (1979), Elzanowski and Epstein (1992), Leon and Epstein (1996), and Münch and Neff (2018). The following chapter is based on Bertram (2015a, b). A fully thermodynamic setting can be found in Bertram (2016). A third-order extension is given by Reiher and Bertram (2018, 2019).[1]

We have already seen in Chap. 1 that the stress power density becomes in the case of a second gradient material (61)

$$\pi_i = 1/\rho(\overset{\langle 2\rangle}{\boldsymbol{T}} \cdot\cdot\, \mathbf{D} + \overset{\langle 3\rangle}{\boldsymbol{T}} \therefore \mathbf{grad\,grad\,v}) \tag{6.1}$$

with a second-order stress tensor $\overset{\langle 2\rangle}{\boldsymbol{T}}$ and a third-order hyperstress tensor $\overset{\langle 3\rangle}{\boldsymbol{T}}$. These stress tensors must be objective under changes of the observer after (62)

$$\overset{\langle 2\rangle}{\boldsymbol{T}}{}^* = \mathbf{Q} * \overset{\langle 2\rangle}{\boldsymbol{T}} \quad \text{and} \quad \overset{\langle 3\rangle}{\boldsymbol{T}}{}^* = \mathbf{Q} * \overset{\langle 3\rangle}{\boldsymbol{T}} \tag{6.2}$$

using the Rayleigh product defined in Chap. 1, Eq. (2), which results here from the mutual rotation between the two observers described by an orthogonal tensor **Q**.

The extension of Cauchy's laws to second-order gradient materials becomes after Chap. 1, Eq. (80)

$$div\ \overset{\langle 2\rangle}{\boldsymbol{T}} - div^2\ \overset{\langle 3\rangle}{\boldsymbol{T}} + \rho\,\mathbf{b} = \rho\,\mathbf{a} \tag{6.3}$$

with the mass density ρ, the specific body force **b** and the acceleration **a**. $\overset{\langle 2\rangle}{\boldsymbol{T}}$ is symmetric due to the balance of moment of momentum. For $\overset{\langle 3\rangle}{\boldsymbol{T}}$, the right subsymmetry can be assumed, since it is work conjugate to *grad grad* **v** which has the same subsymmetry.

For these stress tensors, material laws are needed. In a second gradient theory, one would assume that both of them are functionals of the history or the past process of

- the motion $\mathbf{x} = \boldsymbol{\chi}(\mathbf{x}_0, \tau)$
- the deformation gradient $\mathbf{F} = Grad\,\boldsymbol{\chi}(\mathbf{x}_0, \tau)$
- and the second deformation gradient $Grad^2\boldsymbol{\chi}(\mathbf{x}_0, \tau) = Grad\,\mathbf{F}$

of a material point $\mathbf{x}_0$ with τ being an appropriate time parameter describing the past.

So, we can make the general ansatz with two material functionals

$$\overset{\langle 2\rangle}{\boldsymbol{T}}(\mathbf{x}_0, t) = \mathcal{F}_1(\boldsymbol{\chi}(\mathbf{x}_0, \tau), Grad\,\boldsymbol{\chi}(\mathbf{x}_0, \tau), Grad^2\boldsymbol{\chi}(\mathbf{x}_0, \tau)) \tag{6.4}$$

[1] For more details, see Bertram: *Compendium on Gradient Materials* (various editions).

$$\overset{\langle 3\rangle}{\boldsymbol{T}}(\mathbf{x}_0, t) = \mathcal{F}_2(\boldsymbol{\chi}(\mathbf{x}_0, \tau), Grad\, \boldsymbol{\chi}(\mathbf{x}_0, \tau), Grad^2 \boldsymbol{\chi}(\mathbf{x}_0, \tau)) \tag{6.5}$$

which give the current stresses after some motion.

It is advantageous to submit such material laws to certain invariance requirements to avoid unphysical behaviour. These fundamental requirements are often formulated as *principles of material theory.*

The first and most fundamental of these principles is the

- *Principle of determinism*

In the present context, we will assume it in the following form.

The stresses $\overset{\langle 2\rangle}{\boldsymbol{T}}$ and $\overset{\langle 3\rangle}{\boldsymbol{T}}$ *at a time t in a material point* $\mathbf{x}_0$ *are determined by the history of the motion, the deformation gradient and the second gradient at the same point.*

This principle has already been used by stating the existence of the two functionals (4) and (5).

- *Principle of local action*

The stresses $\overset{\langle 2\rangle}{\boldsymbol{T}}$ and $\overset{\langle 3\rangle}{\boldsymbol{T}}$ *at a material point* $\mathbf{x}_0$ *are determined by the motion of a neighbourhood of that particular point.*

This principle has already been fulfilled in a rather strict way since the knowledge of the motion of *any* open neighbourhood of $\mathbf{x}_0$ already determines all gradients of a smooth function.

- *Principle of Euclidean invariance* (or *objectivity*)

This principle has already been stated in Chap. 1. There, it was shown that the assumption that the power is invariant under changes of observers leads to the existence of *objective* stress tensors (2).

- *Principle of invariance under rigid body modifications*

This principle must be carefully distinguished from the previous one. It states the invariance of the material functionals under a modification of the motion of the form

$$\boldsymbol{\chi}^*(P, \tau) = \mathbf{Q}(\tau) \cdot \boldsymbol{\chi}(P, \tau) + \mathbf{c}(\tau) \tag{6.6}$$

by a time-dependent vector $\mathbf{c}(\tau)$ and a time-dependent proper orthogonal tensor $\mathbf{Q}(\tau)$. This leads to the following transformations of the kinematical variables

- the deformation gradient $\mathbf{F}^* = \mathbf{Q} \cdot \mathbf{F}$
- and the second deformation gradient $(Grad^{\,2}\, \mathbf{F})^* = \mathbf{Q} \cdot Grad^{\,2}\, \mathbf{F}$.

The invariance principle can now be stated in the following form.

The stress laws must be such that

$$\begin{aligned}
&\mathbf{Q}(t) * \mathcal{F}_1(\boldsymbol{\chi}(\mathbf{x}_0, \tau), \mathbf{F}(\mathbf{x}_0, \tau), Grad\, \mathbf{F}(\mathbf{x}_0, \tau)) \\
&\quad = \mathcal{F}_1(\mathbf{Q}(\tau) \cdot \boldsymbol{\chi}(P, \tau) + \mathbf{c}(\tau), \mathbf{Q}(\tau) \cdot \mathbf{F}(\tau), \mathbf{Q}(\tau) \cdot Grad\, \mathbf{F}(\mathbf{x}_0, \tau))
\end{aligned} \tag{6.7}$$

$$\mathbf{Q}(t) * \mathcal{F}_2(\boldsymbol{\chi}(\mathbf{x}_0, \tau), \mathbf{F}(\mathbf{x}_0, \tau), Grad\,\mathbf{F}(\mathbf{x}_0, \tau))$$
$$= \mathcal{F}_2(\mathbf{Q}(\tau) \cdot \boldsymbol{\chi}(P, \tau) + \mathbf{c}(\tau), \mathbf{Q}(\tau) \cdot \mathbf{F}(\tau), \mathbf{Q}(\tau) \cdot Grad\,\mathbf{F}(\mathbf{x}_0, \tau)) \tag{6.8}$$

holds for all proper orthogonal tensors $\mathbf{Q}(\tau)$ *and all vectors* $\mathbf{c}(\tau)$ *and all motions.*

One choice to identically fulfil this principle is by using material variables which are invariant under superimposed rigid body modifications. Material kinematical variables are

- the right Cauchy–Green tensor $\mathbf{C} = \mathbf{F}^T \cdot \mathbf{F}$
- the **curvature triadic**

$$\overset{\langle 3\rangle}{\boldsymbol{K}} := \mathbf{F}^{-1} \cdot Grad\,\mathbf{F} \tag{6.9}$$

and material dynamical variables are

- the second Piola–Kirchhoff tensor

$$\overset{\langle 2\rangle}{\boldsymbol{S}} := \mathbf{F}^{-1} * (J\,\overset{\langle 2\rangle}{\boldsymbol{T}}) \tag{6.10}$$

- and the third-order material hyperstress tensor defined as the pullback of $\overset{\langle 3\rangle}{\boldsymbol{T}}$ to the reference placement

$$\overset{\langle 3\rangle}{\boldsymbol{S}} := \mathbf{F}^{-1} * (J\,\mathbf{F} \cdot \mathbf{F}^T \cdot \overset{\langle 3\rangle}{\boldsymbol{T}}) \tag{6.11}$$

with the Jacobian $J := det(\mathbf{F})$.

These stress tensors are work conjugate to $\mathbf{C}$ and $\overset{\langle 3\rangle}{\boldsymbol{K}}$ since the specific stress power can be written as

$$\pi_i = 1/\rho_0[1/2\,\overset{\langle 2\rangle}{\boldsymbol{S}} \cdot\cdot\, \mathbf{C}^{\bullet} + \overset{\langle 3\rangle}{\boldsymbol{S}} \,\therefore\, \overset{\langle 3\rangle}{\boldsymbol{K}}{}^{\bullet}]. \tag{6.12}$$

The set of all couples like $(\mathbf{C}, \overset{\langle 3\rangle}{\boldsymbol{K}})$ with a positive-definite symmetric tensor in the first entry and a triadic with right subsymmetry in the second forms the **configuration space** of our theory. Dual to it, we have the **stress space** consisting of couples like $(\overset{\langle 2\rangle}{\boldsymbol{S}}, \overset{\langle 3\rangle}{\boldsymbol{S}})$ with a symmetric tensor in the first entry and a triadic with right subsymmetry in the second.

A reduced form of the two stress laws is then

$$\overset{\langle 2\rangle}{S}(\mathbf{x}_0, t) = \mathcal{G}_1(\mathbf{C}(\mathbf{x}_0, \tau), \overset{\langle 3\rangle}{\boldsymbol{K}}(\mathbf{x}_0, \tau)) \tag{6.13}$$

$$\overset{\langle 3\rangle}{S}(\mathbf{x}_0, t) = \mathcal{G}_2(\mathbf{C}(\mathbf{x}_0, \tau), \overset{\langle 3\rangle}{\boldsymbol{K}}(\mathbf{x}_Q, \tau)) \tag{6.14}$$

with two history functionals $\mathcal{G}_1$ and $\mathcal{G}_2$. Of course, this is not the only possible reduced form. In fact, there are infinitely many of them. In the end, the particular choice of them is immaterial, i.e., they all can describe exactly the same material behaviour.

Second Gradient Elasticity

The material laws which have been postulated before remain rather vague. In particular, the dependence on histories or processes has not yet been specified. This, however, is no deficit if we go into elasticity. Here, the history functional becomes a function of the present values of the independent variables. Now, all variables depend on the same point at the same instant of time. Therefore, we can drop the dependence on $(\mathbf{x}_0, t)$ for brevity and simply write for the elastic stress laws

$$\overset{\langle 2\rangle}{\boldsymbol{S}} = k(\mathbf{C}, \overset{\langle 3\rangle}{\boldsymbol{K}}) \tag{6.15}$$

$$\overset{\langle 3\rangle}{\boldsymbol{S}} = K(\mathbf{C}, \overset{\langle 3\rangle}{\boldsymbol{K}}) \tag{6.16}$$

with two functions k and K of the present configuration as the general reduced forms for second gradient elasticity.

A *hyperelastic* second-order gradient material would allow for the introduction of an elastic energy (already as a reduced form) $w(\mathbf{C}, \overset{\langle 3\rangle}{\boldsymbol{K}})$ such that the specific stress power becomes

$$\pi_i = w(\mathbf{C}, \overset{\langle 3\rangle}{\boldsymbol{K}})^{\bullet}. \tag{6.17}$$

By the chain rule and (12), this gives

$$\begin{aligned} &1/\rho_0(1/2)\, \overset{\langle 2\rangle}{\boldsymbol{S}} \cdot\cdot\, \mathbf{C}^{\bullet} + \overset{\langle 3\rangle}{\boldsymbol{S}} \,\therefore\, \overset{\langle 3\rangle}{\boldsymbol{K}}{}^{\bullet}) \\ &= \partial w(\mathbf{C}, \overset{\langle 3\rangle}{\boldsymbol{K}})/\partial \mathbf{C} \cdot\cdot\, \mathbf{C}^{\bullet} + \partial w(\mathbf{C}, \overset{\langle 3\rangle}{\boldsymbol{K}})/\partial \overset{\langle 3\rangle}{\boldsymbol{K}} \,\therefore\, \overset{\langle 3\rangle}{\boldsymbol{K}}{}^{\bullet}. \end{aligned} \tag{6.18}$$

By comparison, we obtain the potential relations for the elastic laws

$$\overset{\langle 2\rangle}{\boldsymbol{S}} = k(\mathbf{C}, \overset{\langle 3\rangle}{\boldsymbol{K}}) = 2\rho_0 \partial w(\mathbf{C}, \overset{\langle 3\rangle}{\boldsymbol{K}})/\partial \mathbf{C} \tag{6.19}$$

$$\overset{\langle 3\rangle}{\boldsymbol{S}} = K(\mathbf{C}, \overset{\langle 3\rangle}{\boldsymbol{K}}) = \rho_0 \partial w(\mathbf{C}, \overset{\langle 3\rangle}{\boldsymbol{K}})/\partial \overset{\langle 3\rangle}{\boldsymbol{K}}\,. \tag{6.20}$$

In the sequel, we will only consider elastic materials that are also hyperelastic.

Of course, one can also use Green's strain tensor

$$\mathbf{E}^{\mathrm{G}} := 1/2(\mathbf{C} - \mathbf{I}) \tag{6.21}$$

instead of the **C** or any other of the infinitely many material strain tensors. If one does so, one obtains

$$k(\mathbf{E}^{\mathrm{G}}, \overset{\langle 3\rangle}{\boldsymbol{K}}) = \rho_0 \, \partial w(\mathbf{E}^{\mathrm{G}}, \overset{\langle 3\rangle}{\boldsymbol{K}})/\partial \mathbf{E}^{\mathrm{G}} \tag{6.22}$$

instead of (19).

Our reduced forms are built by variables which live in the reference placement. The advantage of this fact is that they do not depend on observers and are invariant under translations and rotations of the body. The choice of the reference placement is arbitrary and immaterial. That means that although a change of the reference placement will change all variables like ρ_0, $\mathbf{C}$, $\overset{\langle 3\rangle}{\boldsymbol{K}}$, $\overset{\langle 2\rangle}{\boldsymbol{S}}$, $\overset{\langle 3\rangle}{\boldsymbol{S}}$ as well as the elastic laws w, k and K, it would describe exactly identical material behaviour.

Elastic isomorphy and symmetry. Changes of the reference placement become important when considering material symmetries.[2] We will therefore consider a second reference placement $\underline{\kappa_0}$ indicated by underlining, like all variables with respect to $\underline{\kappa_0}$.

For an arbitrary differentiable field ϕ given in the Lagrangian representation with respect to both reference placements, we obtain by the chain rule

$$\underline{Grad}\,\phi = (Grad\,\phi)\,\mathbf{A} \tag{6.23}$$

where the invertible second-order tensor

$$\mathbf{A} := \underline{Grad}(\kappa_0 \underline{\kappa_0}^{-1}) \tag{6.24}$$

is the gradient of the change of reference placement. It is understood that the field ϕ is defined as a field on the corresponding reference placements, which has identical values for identical material points.

In particular, we find the transformation behaviour of the deformation gradient

$$\underline{\mathbf{F}} = \underline{Grad}\,\chi = \mathbf{F} \cdot \mathbf{A} \tag{6.25}$$

and for its Jacobian

$$\underline{J} = \rho/\underline{\rho_0} = J\,det(\mathbf{A}) = \rho/\rho_0 J_A \tag{6.26}$$

with $J_A := det(\mathbf{A})$, and

[2]See Truesdell and Noll (1965), Sect. 31, p. 76.

$$\underline{\mathbf{C}} = \mathbf{A}^T \cdot \mathbf{C} \cdot \mathbf{A} = \mathbf{A}^T * \mathbf{C}. \tag{6.27}$$

For the second gradient, the transformation is

$$\begin{aligned}\underline{Grad}\,\underline{\mathbf{F}} &= \underline{Grad}(\mathbf{F}\,\mathbf{A}) \\ &= \mathbf{F} \cdot \underline{Grad}\,\mathbf{A} + \mathbf{A}^T * (\mathbf{A}^{-T} \cdot Grad\,\mathbf{F}).\end{aligned} \tag{6.28}$$

Thus,

$$\begin{aligned}\overset{\langle 3\rangle}{\underline{\boldsymbol{K}}} &= \underline{\mathbf{F}}^{-1} \cdot \underline{Grad}\,\mathbf{F} \\ &= \overset{\langle 3\rangle}{\boldsymbol{K}}_{\mathbf{A}} + \mathbf{A}^T * (\mathbf{A}^{-T} \cdot \mathbf{A}^{-1} \cdot \overset{\langle 3\rangle}{\boldsymbol{K}})\end{aligned} \tag{6.29}$$

with the configuration of κ_0 with respect to $\underline{\kappa}_0$

$$\overset{\langle 3\rangle}{\boldsymbol{K}}_{\mathbf{A}} := \mathbf{A}^{-1} \cdot \underline{Grad}\,\mathbf{A}. \tag{6.30}$$

So, the transformation for $\overset{\langle 3\rangle}{\boldsymbol{K}}$ is much more complicated than the one for $\mathbf{C}$. This is due to the property of this triadic which has a tangent (covariant) character in the first entry and a symmetric gradient (contravariant) character in the second and third entry. The same holds in a dual form for the dual hyperstresses as will be seen next.

For the material stresses, we obtain

$$\overset{\langle 2\rangle}{\underline{\boldsymbol{S}}} = \mathbf{A}^{-1} * (J_A\,\overset{\langle 2\rangle}{\boldsymbol{S}}) \tag{6.31}$$

and with (11) for the hyperstresses

$$\overset{\langle 3\rangle}{\underline{\boldsymbol{S}}} = \mathbf{A}^{-1} * (J_A \mathbf{A} \cdot \mathbf{A}^T \cdot \overset{\langle 3\rangle}{\boldsymbol{S}}) \tag{6.32}$$

or inversely

$$\overset{\langle 3\rangle}{\boldsymbol{S}} = \mathbf{A} * (J_A^{-1} \mathbf{A}^{-1} \cdot \mathbf{A}^{-T} \cdot \overset{\langle 3\rangle}{\underline{\boldsymbol{S}}}). \tag{6.33}$$

With these transformations of the variables, we obtain the transformations of the elastic laws

$$k(\kappa_0, \mathbf{C}, \overset{\langle 3\rangle}{\boldsymbol{K}}) = \mathbf{A} * \left[J_A^{-1} k(\underline{\kappa}_0, \mathbf{A}^T * \mathbf{C}, \overset{\langle 3\rangle}{\boldsymbol{K}}_{\mathbf{A}} + \mathbf{A}^T * (\mathbf{A}^{-T} \cdot \mathbf{A}^{-1} \cdot \overset{\langle 3\rangle}{\boldsymbol{K}})) \right] \tag{6.34}$$

$$K(\kappa_0, \mathbf{C}, \overset{\langle 3\rangle}{\boldsymbol{K}}) = \mathbf{A} * (J_A^{-1} \mathbf{A}^{-1} \cdot \mathbf{A}^{-T} \cdot \left[K(\underline{\kappa}_0, \mathbf{A}^T * \mathbf{C}, \overset{\langle 3\rangle}{\boldsymbol{K}}_{\mathbf{A}} + \mathbf{A}^T * (\mathbf{A}^{-T} \cdot \mathbf{A}^{-1} \cdot \overset{\langle 3\rangle}{\boldsymbol{K}})) \right]. \tag{6.35}$$

For the hyperelastic energy introduced in (17), we obtain for the change of the reference placement

$$w(\kappa_0, \mathbf{C}, \overset{\langle 3\rangle}{\boldsymbol{K}}) = w(\underline{\kappa}_0, \mathbf{A}^T * \mathbf{C}, \overset{\langle 3\rangle}{\boldsymbol{K}}_{\mathbf{A}} + \mathbf{A}^T * (\mathbf{A}^{-T} \cdot \mathbf{A}^{-1} \cdot \overset{\langle 3\rangle}{\boldsymbol{K}})). \tag{6.36}$$

So if we write down the elastic laws for a certain material point with respect to two different reference placements, the form of these laws will look rather different, although they describe exactly the same behaviour.

On the other hand, it may happen that two material points also have the same elastic behaviour, even if their elastic laws look different.

In such cases, the two tensors $\mathbf{A}$ and $\overset{\langle 3\rangle}{\boldsymbol{K}}_{\mathbf{A}}$ must exist such that one can transform the elastic laws by the above equations into one another. If we restrict the consideration to reference placements with the same density, the tensors $\mathbf{A}$ are unimodular (*det* $\mathbf{A}$ $= 1$). In such cases, we call the two points elastically or materially **isomorphic**.

A **symmetry transformation** is an automorphism or an isomorphism w.r.t. the same point. More precisely, we call the tensor couple $(\mathbf{A}, \overset{\langle 3\rangle}{\boldsymbol{K}}_{\mathbf{A}})$ a symmetry transformation of the elastic law $w(\mathbf{C}, \overset{\langle 3\rangle}{\boldsymbol{K}})$ if

$$w(\mathbf{C}, \overset{\langle 3\rangle}{\boldsymbol{K}}) = w(\mathbf{A}^T * \mathbf{C}, \overset{\langle 3\rangle}{\boldsymbol{K}}_{\mathbf{A}} + \mathbf{A}^T * (\mathbf{A}^{-T} \cdot \mathbf{A}^{-1} \cdot \overset{\langle 3\rangle}{\boldsymbol{K}})) \tag{6.37}$$

holds for all configurations $(\mathbf{C}, \overset{\langle 3\rangle}{\boldsymbol{K}})$.

All symmetry transformations form a group under composition, the *symmetry group* of the energy. This group characterizes the type of elasticity and helps us to give the material laws a more specific form by use of representations of the corresponding group.

While the role of the second-order tensor $\mathbf{A}$ is familiar to us from the context of simple materials, rather little is known up to now about the role of the triadic $\overset{\langle 3\rangle}{\boldsymbol{K}}_{\mathbf{A}}$ in the symmetry transformation.

We can call an elastic material for which the symmetry group contains all symmetry transformations of the form $(\mathbf{Q}, \overset{\langle 3\rangle}{\boldsymbol{O}})$ with an orthogonal $\mathbf{Q}$ and the zero triadic $\overset{\langle 3\rangle}{\boldsymbol{O}}$ *isotropic*. If this holds only for proper orthogonal tensors $\mathbf{Q}$, it would be *hemitropic*. While the sign of the orthogonal tensor cancels out for simple materials, this is no longer the case for second gradient materials, so we have to carefully distinguish between isotropy and hemitropy.

For such orthogonal symmetry transformations, the equations become simpler

$$w(\mathbf{C}, \overset{\langle 3\rangle}{\boldsymbol{K}}) = w(\underline{\kappa}_0, \mathbf{Q} * \mathbf{C}, \mathbf{Q} * \overset{\langle 3\rangle}{\boldsymbol{K}}) \tag{6.38}$$

$$\mathbf{Q} * k(\mathbf{C}, \overset{\langle 3\rangle}{\boldsymbol{K}}) = k(\mathbf{Q} * \mathbf{C}, \mathbf{Q} * \overset{\langle 3\rangle}{\boldsymbol{K}}) \tag{6.39}$$

$$\mathbf{Q} * K(\mathbf{C}, \overset{\langle 3\rangle}{\boldsymbol{K}}) = K(\mathbf{Q} * \mathbf{C}, \mathbf{Q} * \overset{\langle 3\rangle}{\boldsymbol{K}}). \tag{6.40}$$

Such functions are called *hemitropic tensor functions* if they hold for all proper orthogonal tensors, or *isotropic tensor functions* if they hold for all orthogonal tensors.

Finite linear second-order elastic materials. The elastic materials that we have introduced in the preceding section are generally nonlinear functions of the kinematical variables. One obtains a much more particular form by limiting to the linear case. In what follows, we will give such linear laws. However, one must keep in mind that in the context of finite deformations the linear laws depend on the choice of the variables. And since there are infinitely many of them, there are also infinitely many linear laws. We already know this fact from the classical St. Venant–Kirchhoff law.

Instead of using the right Cauchy–Green strain tensors $\mathbf{C}$, we will linearize the Green strain tensor $\mathbf{E}^{\mathrm{G}}$ which is zero in the reference placement. So, our linear law applies to situations where the material point is in a configuration not far from the reference one. This means that both $\mathbf{E}^{\mathrm{G}}$ and $\overset{\langle 3\rangle}{\boldsymbol{K}}$ are close to the zero dyadic and zero triadic, respectively, or that the norms of them are small.

One obtains a finite linear elastic theory by introducing the energy as a square form

$$\rho_0 w(\mathbf{E}^{\mathrm{G}}, \overset{\langle 3\rangle}{\boldsymbol{K}}) = 1/2\, \mathbf{E}^{\mathrm{G}} \cdot\cdot \overset{\langle 4\rangle}{\boldsymbol{E}}_{22} \cdot\cdot \mathbf{E}^{\mathrm{G}} + \mathbf{E}^{\mathrm{G}} \cdot\cdot \overset{\langle 5\rangle}{\boldsymbol{E}}_{23} \therefore \overset{\langle 3\rangle}{\boldsymbol{K}} + 1/2\, \overset{\langle 3\rangle}{\boldsymbol{K}} \therefore \overset{\langle 6\rangle}{\boldsymbol{E}}_{33} \therefore \overset{\langle 3\rangle}{\boldsymbol{K}} \tag{6.41}$$

with three elasticity tensors $\overset{\langle 4\rangle}{\boldsymbol{E}}_{22}$, $\overset{\langle 5\rangle}{\boldsymbol{E}}_{23}$ and $\overset{\langle 6\rangle}{\boldsymbol{E}}_{33}$.

Formally, a square form of a second- and a third-order tensor has $(9 + 27)^2 = 1296$ constants. Because of the symmetry of $\mathbf{E}^{\mathrm{G}}$ and the subsymmetry of $\overset{\langle 3\rangle}{\boldsymbol{K}}$, this can be reduced to $(6 + 18)^2 = 576$. Because of the symmetry of the square form, only 300 independent constants remain.

The stress laws become after (20) and (22)

$$k(\mathbf{E}^{G}, \overset{\langle 3\rangle}{\boldsymbol{K}}) = \overset{\langle 4\rangle}{\boldsymbol{E}}_{22} \cdot\cdot \mathbf{E}^{G} + \overset{\langle 5\rangle}{\boldsymbol{E}}_{23} \therefore \overset{\langle 3\rangle}{\boldsymbol{K}} \tag{6.42}$$

$$K(\mathbf{E}^{\mathrm{G}}, \overset{\langle 3\rangle}{\boldsymbol{K}}) = \mathbf{E}^{\mathrm{G}} \cdot\cdot \overset{\langle 5\rangle}{\boldsymbol{E}}_{23} + \overset{\langle 6\rangle}{\boldsymbol{E}}_{33} \therefore \overset{\langle 3\rangle}{\boldsymbol{K}}\,. \tag{6.43}$$

If the material is hemitropic, these elasticity tensors must also be hemitropic, i.e.

$$\overset{\langle 4\rangle}{\boldsymbol{E}}_{22} = \mathbf{Q} * \overset{\langle 4\rangle}{\boldsymbol{E}}_{22}$$
$$\overset{\langle 5\rangle}{\boldsymbol{E}}_{23} = \mathbf{Q} * \overset{\langle 5\rangle}{\boldsymbol{E}}_{23}$$

$$\overset{\langle 6\rangle}{\boldsymbol{E}}_{33} = \mathbf{Q} * \overset{\langle 6\rangle}{\boldsymbol{E}}_{33} \tag{6.44}$$

for all proper orthogonal tensors $\mathbf{Q}$.

In our case, we can choose the following complete representation for these hemitropic tensors listed in, e.g., Bertram (2019)

$$\begin{aligned}
\overset{\langle 4\rangle}{\boldsymbol{E}}_{22} &= \alpha_1 \mathbf{I} \otimes \mathbf{I} + \alpha_2 \overset{\langle 4\rangle}{\boldsymbol{I}} \\
\overset{\langle 4\rangle}{\boldsymbol{E}}_{22} &= \alpha_3 \varepsilon_{ijk} \mathbf{e}_i \otimes \mathbf{e}_l \otimes \mathbf{e}_j \otimes \mathbf{e}_l \otimes \mathbf{e}_k \\
\overset{\langle 4\rangle}{\boldsymbol{E}}_{22} &= \alpha_4 \overset{\langle 6\rangle}{\boldsymbol{I}} + \alpha_5/2(\overset{\langle 6\rangle}{\boldsymbol{I}}{}^{[12]} + \overset{\langle 6\rangle}{\boldsymbol{I}}{}^{[13]}) \\
&\quad + \alpha_6 \mathbf{e}_i \otimes \mathbf{I} \otimes \mathbf{e}_i \otimes \mathbf{I} + \alpha_7 \mathbf{e}_i \otimes \mathbf{I} \otimes \mathbf{I} \otimes \mathbf{e}_i + \alpha_8 \mathbf{I} \otimes \mathbf{e}_i \otimes \mathbf{I} \otimes \mathbf{e}_i
\end{aligned} \tag{6.45}$$

with respect to an orthonormal vector basis $\{\mathbf{e}_i\}$ for any scalar factors α_1 to α_8. Here, $\overset{\langle 6\rangle}{\boldsymbol{I}}{}^{[12]}$ stands for the symmetrizer in the first and second entry. Note that any linear combination of hemitropic tensors is also hemitropic.

The general hemitropic square form is therefore

$$\begin{aligned}
\rho_0 w(\mathbf{E}^{\mathrm{G}}, \overset{\langle 3\rangle}{\boldsymbol{K}}) &= \alpha_1/2\, tr^2 \mathbf{E}^{\mathrm{G}} + \alpha_2/2 \mathbf{E}^{\mathrm{G}} \cdot\cdot \mathbf{E}^{\mathrm{G}} \\
&+ \alpha_3 \mathbf{E}^{\mathrm{G}} \cdot\cdot (\boldsymbol{\varepsilon} \cdot\cdot \overset{\langle 3\rangle}{\boldsymbol{K}}) \\
&+ \alpha_4/2 \overset{\langle 3\rangle}{\boldsymbol{K}} \,\therefore\, \overset{\langle 3\rangle}{\boldsymbol{K}} \\
&+ \alpha_5/4 \overset{\langle 3\rangle}{\boldsymbol{K}} \,\therefore\, (\overset{\langle 3\rangle}{\boldsymbol{K}}{}^{[12]} + \overset{\langle 3\rangle}{\boldsymbol{K}}{}^{[13]}) \\
&+ \alpha_6/2 (\overset{\langle 3\rangle}{\boldsymbol{K}} \cdot\cdot \mathbf{I}) \cdot (\overset{\langle 3\rangle}{\boldsymbol{K}} \cdot\cdot \mathbf{I}) \\
&+ \alpha_7/2 (\mathbf{I} \cdot\cdot \overset{\langle 3\rangle}{\boldsymbol{K}}) \cdot (\overset{\langle 3\rangle}{\boldsymbol{K}} \cdot\cdot \mathbf{I}) \\
&+ \alpha_8/2 (\mathbf{I} \cdot\cdot \overset{\langle 3\rangle}{\boldsymbol{K}}) \cdot (\mathbf{I} \cdot\cdot \overset{\langle 3\rangle}{\boldsymbol{K}})
\end{aligned} \tag{6.46}$$

with only eight material constants. α_1 and α_2 correspond to the classical Lamé constants.

If the material is isotropic, then the mixed term with α_3 vanishes.

The resulting stresses are after (19)

$$\overset{\langle 2\rangle}{\boldsymbol{S}} = \alpha_1 (tr\, \mathbf{E}^{\mathrm{G}}) \mathbf{I} + \alpha_2 \mathbf{E}^{\mathrm{G}} + \alpha_3 sym(\boldsymbol{\varepsilon} \cdot\cdot \overset{\langle 3\rangle}{\boldsymbol{K}}) \tag{6.47}$$

and after (20)

$$\overset{\langle 3\rangle}{\boldsymbol{S}} = sym^{[23]} \Big[\alpha_3 \boldsymbol{\varepsilon} \cdot \mathbf{E}^{\mathrm{G}} + \alpha_4 \overset{\langle 3\rangle}{\boldsymbol{K}}$$

$$+\alpha_5 \overset{\langle 6\rangle}{\boldsymbol{I}}^{[13]} \therefore \overset{\langle 3\rangle}{\boldsymbol{K}} + \alpha_6 \overset{\langle 3\rangle}{\boldsymbol{K}} \cdot\cdot \mathbf{I} \otimes \mathbf{I}$$

$$\left. + \alpha_7 \mathbf{I} \otimes \overset{\langle 3\rangle}{\boldsymbol{K}} \cdot\cdot \mathbf{I}) + \alpha_8 \mathbf{I} \otimes \mathbf{I} \cdot\cdot \overset{\langle 3\rangle}{\boldsymbol{K}} \right] . \qquad (6.48)$$

Finite Elastoplasticity

The literature on finite gradient plasticity is rather sparse. Examples are Gurtin and Anand (2005), Gurtin et al. (2009), Gurtin (2010), Luscher et al. (2010) and Cleja-Ťigoiu (2002, 2010, 2013). Our present approach, however, is different from theirs. It is based on Bertram (2015a, b). A fully thermodynamic setting can be found in Bertram (2016). A third-order extension is given by Reiher and Bertram (2018, 2019).[3]

The basic concept of elastoplasticity is the existence of an elastic range at any state that the material happens to be currently in. We will introduce this concept in the configuration space in the following way.

Let $(\mathbf{C}, \overset{\langle 3\rangle}{\boldsymbol{K}})$ be the current configuration of an elastoplastic material. Then, we assume that there exists an associated **elastic range** consisting of the following two ingredients

1. a bounded neighbourhood $\mathcal{E}_p$ of this configuration
2. an elastic energy w_p defined on $\mathcal{E}_p$

such that for any kinematical process starting at $(\mathbf{C}, \overset{\langle 3\rangle}{\boldsymbol{K}})$ which remains entirely in $\mathcal{E}_p$ the stresses are given by w_p through the potentials (19) and (20). The elastic behaviour can be isotropic or anisotropic, and linear or nonlinear.

We will call such configuration process segments that remain entirely in the current elastic range, **elastic processes**. In what follows, we will give all variables which are linked to a particular elastic range and are not altered by any elastic processes a name with p.

If we prescribe locally a configuration process, it is elastic as long as it remains in the current elastic range. However, if it eventually crosses the boundary of this elastic range, the material has to necessarily pass through other elastic ranges. Such process segments are no longer elastic but plastic or states of **yielding**. Note that yielding can only occur if the configuration is on the boundary of the current $\mathcal{E}_p$ and about to cross it. These conditions will later be practically controlled by a yield and a loading criterion.

In such a way, we can decompose any configuration process of our elastoplastic material point into

- elastic segments, where the associated elastic range remains frozen, and

[3]For more details, see Bertram: Compendium on Gradient Materials (various editions).

- plastic segments, where it continuously passes through different elastic ranges, and the current stresses are determined by the energy w_p of the particular current elastic range.

In the latter case of yielding, both $\mathcal{E}_p$ and w_p will evolve. To model these evolutions will be our next task.

Isomorphy of Elastic Ranges. The evolution of the elastic energy during yielding can happen in many different ways. If the material possesses a microstructure like a crystallographic one, the elastic behaviour will be affected by the evolution of this structure. This may lead to changes of the elastic stiffness as it can be observed in rolling and metal forming. However, to model such changes is extremely difficult.

The most simple assumption would be that the energy remains identical during yielding, in the sense previously described as isomorphic elastic behaviour. This concept can be given both a precise mathematical form and clear physical interpretation; see Bertram (2005). Its application, however, is limited, since only some materials show this property like crystals.

To make the assumption of identical elastic behaviour more precise, we consider two different elastic ranges of the same material point and write down the isomorphy condition between them in the form

$$w_{p1}(\mathbf{C}, \overset{\langle 3\rangle}{\boldsymbol{K}}) = w_{p2}(\mathbf{P}^T * \mathbf{C}, \overset{\langle 3\rangle}{\boldsymbol{P}} + \mathbf{P}^T * (\mathbf{P}^{-T} \cdot \mathbf{P}^{-1} \cdot \overset{\langle 3\rangle}{\boldsymbol{K}})). \tag{6.49}$$

The isomorphy transformation is determined by a second-order unimodular tensor $\mathbf{P}$ and a triadic $\overset{\langle 3\rangle}{\boldsymbol{P}}$ with right subsymmetry. These tensors are internal variables in this theory and need evolution equations. The stress laws become after (34) and (35)

$$k_{p1}(\mathbf{C}, \overset{\langle 3\rangle}{\boldsymbol{K}}) = \mathbf{P} * \left[k_{p2}(\mathbf{P}^T * \mathbf{C}, \overset{\langle 3\rangle}{\boldsymbol{P}} + \mathbf{P}^T * (\mathbf{P}^{-T} \cdot \mathbf{P}^{-1} \cdot \overset{\langle 3\rangle}{\boldsymbol{K}})) \right] \tag{6.50}$$

$$K_{p1}(\mathbf{C}_2, \overset{\langle 3\rangle}{\boldsymbol{K}}) = \mathbf{P} * \left[\mathbf{P}^{-T} \cdot \mathbf{P}^{-1} \cdot K_{p2}(\mathbf{P}^T * \mathbf{C}, \overset{\langle 3\rangle}{\boldsymbol{P}} + \mathbf{P}^T * (\mathbf{P}^{-T} \cdot \mathbf{P}^{-1} \cdot \overset{\langle 3\rangle}{\boldsymbol{K}}) \right]. \tag{6.51}$$

Practically, one would choose an elastic reference energy w_0 which is constant in time and is isomorphic to the current energy, which is not constant. Thus,

$$w_p(\mathbf{C}, \overset{\langle 3\rangle}{\boldsymbol{K}}) = w_0(\mathbf{P}^T * \mathbf{C}, \overset{\langle 3\rangle}{\boldsymbol{P}} + \mathbf{P}^T * (\mathbf{P}^{-T} \cdot \mathbf{P}^{-1} \cdot \overset{\langle 3\rangle}{\boldsymbol{K}})) \tag{6.52}$$

holds for each elastic range.

The elastic laws are then given as the potentials by

$$\overset{\langle 2\rangle}{\boldsymbol{S}} = \mathbf{P} * \left[2\rho_0 \partial w_0(\mathbf{C}_e, \overset{\langle 3\rangle}{\boldsymbol{K}}_e)/\partial \mathbf{C}_e \right] \tag{6.53}$$

$$\overset{\langle 3\rangle}{\boldsymbol{S}} = \mathbf{P} * \left\{ \rho_0 \mathbf{P}^{-1} \cdot \mathbf{P}^{-T} \cdot [\partial w_0(\mathbf{C}_e, \overset{\langle 3\rangle}{\boldsymbol{K}}_e)/\partial \overset{\langle 3\rangle}{\boldsymbol{K}}_e \right\} \tag{6.54}$$

with the abbreviations

$$\mathbf{C}_e := \mathbf{P}^T \cdot \mathbf{C} \cdot \mathbf{P} = \mathbf{P}^T * \mathbf{C} \tag{6.55}$$

$$\overset{\langle 3\rangle}{\boldsymbol{K}}_e := \mathbf{P}^T * (\mathbf{P}^{-T} \cdot \mathbf{P}^{-1} \cdot \overset{\langle 3\rangle}{\boldsymbol{K}}) + \overset{\langle 3\rangle}{\boldsymbol{P}} \,. \tag{6.56}$$

Yield Criterion. We will now consider the evolution of the set $\mathcal{E}_p$ during yielding. The easiest way to describe the elastic range as a subset of the configuration space is to introduce a yield criterion as a level set function.

For this purpose, we decompose the set $\mathcal{E}_p$ topologically into its interior $\mathcal{E}_p^o$ and its boundary $\partial\mathcal{E}_p$. The latter is called **yield surface** (in the configuration space). By the elastic laws (53) and (54), one can likewise transform the yield surface into the stress space if preferred.

Next, we introduce a real-valued tensor function Φ_p on the configuration space which assigns to each configuration $(\mathbf{C}, \overset{\langle 3\rangle}{\boldsymbol{K}})$ a real number in such a way that the kernel of this function coincides with the yield limit

$$\Phi_p(\mathbf{C}, \overset{\langle 3\rangle}{\boldsymbol{K}}) = 0 \quad \Leftrightarrow \quad (\mathbf{C}, \overset{\langle 3\rangle}{\boldsymbol{K}}) \in \partial\mathcal{E}_p \tag{6.57}$$

For distinguishing points in the interior and in the exterior of the elastic ranges, we postulate

$$\Phi_R(\mathbf{C}, \overset{\langle 3\rangle}{\boldsymbol{K}}) < 0 \quad \Leftrightarrow \quad (\mathbf{C}, \overset{\langle 3\rangle}{\boldsymbol{K}}) \in \mathcal{E}_p^o \tag{6.58}$$

and, consequently,

$$\Phi_p(\mathbf{C}, \overset{\langle 3\rangle}{\boldsymbol{K}}) > 0 \quad \Leftrightarrow \quad \text{outside of} \, \mathcal{E}_p. \tag{6.59}$$

So, the function Φ_p works as a level set or indicator function for $\mathcal{E}_p$. It will always exist but never be unique. We refer to it as a *yield criterion* since yielding can only occur if the following two conditions are fulfilled

(i) the configuration is on the current yield surface $\partial\mathcal{E}_p$; i.e. the **yield condition** is fulfilled

$$\Phi_p(\mathbf{C}, \overset{\langle 3\rangle}{\boldsymbol{K}}) = 0 \tag{6.60}$$

(ii) and the process is such that the current configuration is about to cross the yield limit

$$\Phi_p^* = \partial\Phi_p/\partial\mathbf{C} \cdot\cdot\, \mathbf{C}^\bullet + \partial\Phi_p/\partial \overset{\langle 3\rangle}{\boldsymbol{K}} \,\therefore\, \overset{\langle 3\rangle}{\boldsymbol{K}}{}^\bullet > 0 \tag{6.61}$$

which is called **loading condition**.

This function Φ_p is part of the particular elastic range. So, it will evolve if the set $\mathcal{E}_p$ evolves. To describe this function for all elastic ranges in the same form, we introduce a number of parameters here notated as (but not limited to) a tensor $\mathbf{H}_p$ such that the general form of the yield criterion becomes

$$\varphi(\mathbf{P}, \overset{\langle 3\rangle}{\boldsymbol{P}}, \mathbf{C}, \overset{\langle 3\rangle}{\boldsymbol{K}}, \mathbf{H}_p) \tag{6.62}$$

thus depending on all variables, such that

$$\Phi_p(\mathbf{C}, \overset{\langle 3\rangle}{\boldsymbol{K}}) = \varphi(\mathbf{P}, \overset{\langle 3\rangle}{\boldsymbol{P}}, \mathbf{C}, \overset{\langle 3\rangle}{\boldsymbol{K}}, \mathbf{H}_p) \tag{6.63}$$

holds for each particular elastic range. The variables $\mathbf{H}_p$ will later be interpreted as hardening/softening variables for which a hardening rule will be needed.

With this extension, we obtain for the yield condition (60)

$$\varphi(\mathbf{P}, \overset{\langle 3\rangle}{\boldsymbol{P}}, \mathbf{C}, \overset{\langle 3\rangle}{\boldsymbol{K}}, \mathbf{H}_p) = 0 \tag{6.64}$$

and for the loading condition (61)

$$\partial\varphi/\partial\mathbf{C} \cdot\cdot\, \mathbf{C}^\bullet + \partial\varphi/\partial \overset{\langle 3\rangle}{\boldsymbol{K}} \,\therefore\, \overset{\langle 3\rangle}{\boldsymbol{K}}{}^\bullet > 0 \tag{6.65}$$

where the plastic variables are kept constant.

Flow and Hardening Rules. For the evolution of the internal plastic variables, evolution equations are needed, namely two **flow rules** for $\mathbf{P}$ and $\overset{\langle 3\rangle}{\boldsymbol{P}}$ and a **hardening rule** for $\mathbf{H}_p$. A general ansatz for them is to let their increments depend on the increments of the independent variables and on the state of all internal variables

$$\begin{aligned}
\mathbf{P}^\bullet &= f(\mathbf{P}, \overset{\langle 3\rangle}{\boldsymbol{P}}, \mathbf{C}, \overset{\langle 3\rangle}{\boldsymbol{K}}, \mathbf{H}_p, \mathbf{C}^*, \overset{\langle 3\rangle}{\boldsymbol{K}}{}^\bullet) \\
\overset{\langle 3\rangle}{\boldsymbol{P}}{}^\bullet &= F(\mathbf{P}, \overset{\langle 3\rangle}{\boldsymbol{P}}, \mathbf{C}, \overset{\langle 3\rangle}{\boldsymbol{K}}, \mathbf{H}_p, \mathbf{C}^\bullet, \overset{\langle 3\rangle}{\boldsymbol{K}}{}^\bullet) \\
\mathbf{H}_p^\bullet &= h(\mathbf{P}, \overset{\langle 3\rangle}{\boldsymbol{P}}, \mathbf{C}, \overset{\langle 3\rangle}{\boldsymbol{K}}, \mathbf{H}_p, \mathbf{C}, \overset{\langle 3\rangle}{\boldsymbol{K}}{}^\bullet)
\end{aligned} \tag{6.66}$$

in the form of rate-independent ODEs f, F and h as customary in plasticity.

The rate independence can be assured in the usual way by the introduction of a **plastic consistency parameter** $\lambda \geq 0$

$$\mathbf{P}^{\bullet} = \lambda f^{\circ}(\mathbf{P}, \overset{\langle 3\rangle}{\boldsymbol{P}}, \mathbf{C}, \overset{\langle 3\rangle}{\boldsymbol{K}}, \mathbf{H}_p, \mathbf{C}^{\circ}, \overset{\langle 3\rangle}{\boldsymbol{K}}{}^{\circ})$$

$$\overset{\langle 3\rangle}{\boldsymbol{P}}{}^{\bullet} = \lambda F^{\circ}(\mathbf{P}, \overset{\langle 3\rangle}{\boldsymbol{P}}, \mathbf{C}, \overset{\langle 3\rangle}{\boldsymbol{K}}, \mathbf{H}_p, \mathbf{C}^{\circ}, \overset{\langle 3\rangle^{\circ}}{\boldsymbol{K}})$$

$$\mathbf{H}_p^{\bullet} = \lambda h^{\circ}(\mathbf{P}, \overset{\langle 3\rangle}{\boldsymbol{P}}, \mathbf{C}, \overset{\langle 3\rangle}{\boldsymbol{K}}, \mathbf{H}_p, \mathbf{C}^{\circ}, \overset{\langle 3\rangle}{\boldsymbol{K}}{}^{\circ}) \tag{6.67}$$

where we normed the increments of the kinematical variables

$$\mathbf{C}^{\circ} := \mathbf{C}^{\bullet}/\mu \quad \text{and} \quad \overset{\langle 3\rangle}{\boldsymbol{K}}{}^{\circ} := \overset{\langle 3\rangle}{\boldsymbol{K}}{}^{\bullet}/\mu \tag{6.68}$$

by a factor

$$\mu := \sqrt{(|\mathbf{C}^{\bullet}|^2 + L^2 \left|\overset{\langle 3\rangle}{\boldsymbol{K}}{}^{\bullet}\right|^2)} \tag{6.69}$$

which is positive during yielding. The positive constant L with the dimension of a length is necessary for dimensional reasons and controls the ratio of yielding due to $\mathbf{C}^{\bullet}$ and $\overset{\langle 3\rangle}{\boldsymbol{K}}{}^{\bullet}$, respectively.

The three functions f°, F° and h° determine the directions of the flow and hardening, while the amount is finally given by the consistency parameter. The consistency parameter is zero during elastic processes. During yielding, it can be calculated by the yield condition (64)

$$\begin{aligned}
0 &= \varphi(\mathbf{P}, \overset{\langle 3\rangle}{\boldsymbol{P}}, \mathbf{C}, \overset{\langle 3\rangle}{\boldsymbol{K}}, \mathbf{H}_p)^{\bullet} \\
&= \partial\varphi/\partial\mathbf{P} \cdot\cdot\, \mathbf{P}^{\bullet} + \partial\varphi/\partial \overset{\langle 3\rangle}{\boldsymbol{P}} \therefore \overset{\langle 3\rangle}{\boldsymbol{P}}{}^{\bullet} + \partial\varphi/\partial\mathbf{C} \cdot\cdot\, \mathbf{C}^{\bullet} \\
&\quad + \partial\varphi/\partial \overset{\langle 3\rangle}{\boldsymbol{K}} \therefore \overset{\langle 3\rangle}{\boldsymbol{K}}{}^{\bullet} + \partial\varphi/\partial\mathbf{H}_p \cdot\cdot\, \mathbf{H}_p^{\bullet} \\
&= \partial\varphi/\partial\mathbf{P} \cdot\cdot\, \lambda f^{\circ}(\mathbf{P}, \overset{\langle 3\rangle}{\boldsymbol{P}}, \mathbf{C}, \overset{\langle 3\rangle}{\boldsymbol{K}}, \mathbf{H}_p, \mathbf{C}^{\circ}, \overset{\langle 3\rangle}{\boldsymbol{K}}{}^{\circ}) \\
&\quad + \partial\varphi/\partial \overset{\langle 3\rangle}{\boldsymbol{P}} \therefore \lambda F^{\circ}(\mathbf{P}, \overset{\langle 3\rangle}{\boldsymbol{P}}, \mathbf{C}, \overset{\langle 3\rangle}{\boldsymbol{K}}, \mathbf{H}_p, \mathbf{C}^{\circ}, \overset{\langle 3\rangle}{\boldsymbol{K}}{}^{\circ}) + \partial\varphi/\partial\mathbf{C} \cdot\cdot\, \mathbf{C}^{\bullet} \\
&\quad + \partial\varphi/\partial \overset{\langle 3\rangle}{\boldsymbol{K}} \therefore \overset{\langle 3\rangle}{\boldsymbol{K}}{}^{\bullet} + \partial\varphi/\partial\mathbf{H}_p \cdot\cdot\, \lambda h^{\circ}(\mathbf{P}, \overset{\langle 3\rangle}{\boldsymbol{P}}, \mathbf{C}, \overset{\langle 3\rangle}{\boldsymbol{K}}, \mathbf{H}_p, \mathbf{C}^{\circ}, \overset{\langle 3\rangle}{\boldsymbol{K}}{}^{\circ})
\end{aligned} \tag{6.70}$$

which gives the quotient

$$\lambda = -\left[\partial\varphi/\partial\mathbf{C} \cdot\cdot\, \mathbf{C}^{\bullet} + \partial\varphi/\partial \overset{\langle 3\rangle}{\boldsymbol{K}} \therefore \overset{\langle 3\rangle}{\boldsymbol{K}}{}^{\bullet}\right] /$$

$$\left[\partial\varphi/\partial\mathbf{P} \cdot\cdot f^{\circ}(\mathbf{P}, \overset{\langle 3\rangle}{\boldsymbol{P}}, \mathbf{C}, \overset{\langle 3\rangle}{\boldsymbol{K}}, \mathbf{H}_p, \mathbf{C}^{\circ}, \overset{\langle 3\rangle}{\boldsymbol{K}}{}^{\circ})\right.$$

$$+\,\partial\varphi/\partial\overset{\langle 3\rangle}{\boldsymbol{P}}\therefore F^{\circ}(\mathbf{P},\overset{\langle 3\rangle}{\boldsymbol{P}},\mathbf{C},\mathbf{K},\mathbf{H}_p,\mathbf{C}^{\circ},\overset{\langle 3\rangle}{\boldsymbol{K}^{\circ}})$$
$$\left.+\partial\varphi/\partial\mathbf{H}_p\cdot\cdot\, h^{\circ}(\mathbf{P},\overset{\langle 3\rangle}{\boldsymbol{P}},\mathbf{C},\overset{\langle 3\rangle}{\boldsymbol{K}},\mathbf{H}_p,\mathbf{C}^{\circ},\overset{\langle 3\rangle}{\boldsymbol{K}^{\circ}})\right]. \tag{6.71}$$

Both numerator and denominator of this ratio are always negative during yielding as a consequence of the loading condition (65), and thus, λ is positive in this case.

If we substitute this value of λ into (67), we obtain the *consistent flow* and *hardening rules*. In all cases (elastic *and* plastic), the Kuhn–Tucker condition

$$\lambda\,\varphi = 0 \quad \text{with} \quad \lambda \geq 0 \quad \text{and} \quad \varphi \leq 0 \tag{6.72}$$

holds since at any time one of the two factors is zero.

With these rules, our model of a second-order elastoplastic material is completed, in a rather general form.

Decomposition of the Stress Power. We will next consider the stress power again and specify it for our elastoplastic material. The specific stress power (18) is by using (53) and (54)

$$\pi_l = I/\rho_0\left\{1/2k_0(\mathbf{C}_e,\overset{\langle 3\rangle}{\boldsymbol{K}_e})\cdot\cdot\,(\mathbf{P}^T * \mathbf{C}^{\bullet})\right.$$
$$\left.+K_0(\mathbf{C}_e,\overset{\langle 3\rangle}{\boldsymbol{K}_e})\therefore\left[\mathbf{P}^T * (\mathbf{P}^{-T}\cdot\mathbf{P}^{-1}\cdot\overset{\langle 3\rangle}{\boldsymbol{K}^{\bullet}})\right]\right\}. \tag{6.73}$$

(55) and (56) give for the rates

$$\mathbf{C}_e^{\bullet} = (\mathbf{P}^T\cdot\mathbf{C}\cdot\mathbf{P})^{\bullet}$$
$$= \mathbf{P}^T * \mathbf{C}^{\bullet} + 2\,sym(\mathbf{C}_e\cdot\mathbf{P}^{-1}\cdot\mathbf{P}^{\bullet}) \tag{6.74}$$

where *sym* stands for the symmetric part, and

$$\overset{\langle 3\rangle}{\boldsymbol{K}}{}^{\bullet\,v}_{\;e} = [\mathbf{P}^T * (\mathbf{P}^{-T}\cdot\mathbf{P}^{-1}\cdot\overset{\langle 3\rangle}{\boldsymbol{K}}) + \overset{\langle 3\rangle}{\boldsymbol{P}}]^{\bullet}$$
$$= \mathbf{P}^T * (\mathbf{P}^{-T}\cdot\mathbf{P}^{-1}\cdot\overset{\langle 3\rangle}{\boldsymbol{K}}\cdot) + \overset{\langle 3\rangle}{\boldsymbol{P}^{\bullet}} - \mathbf{P}^{-1}\cdot\mathbf{P}^{\bullet}\cdot(\overset{\langle 3\rangle}{\boldsymbol{K}_e} - \mathbf{P})$$
$$+\,2\,subsym[(\overset{\langle 3\rangle}{\boldsymbol{K}_e} - \overline{\mathbf{P}})\cdot\mathbf{P}^{-1}\cdot\mathbf{P}^{\bullet}] \tag{6.75}$$

the term with *subsym* being the symmetric part with respect to the right subsymmetry. We substitute this into (73) to obtain

$$\pi_i = I/\rho_0\left\{1/2k_0(\mathbf{C}_e,\overset{\langle 3\rangle}{\boldsymbol{K}_e})\cdot\cdot\,\left[\mathbf{C}_e^{\bullet} - 2\,sym(\mathbf{C}_e\cdot\mathbf{P}^{-1}\cdot\mathbf{P}^{\bullet})\right]\right.$$

$$
\begin{aligned}
&+ K_0(\mathbf{C}_e, \overset{\langle 3\rangle}{\boldsymbol{K}}_e) \therefore \left[\overset{\langle 3\rangle}{\boldsymbol{K}}{}^{\bullet}_e - \overset{\langle 3\rangle}{\boldsymbol{P}}{}^{\bullet} + \mathbf{P}^{-1} \cdot \mathbf{P}^{\bullet}(\overset{\langle 3\rangle}{\boldsymbol{K}}_e - \overset{\langle 3\rangle}{\boldsymbol{P}}) \right. \\
&\left. -2\, subsym\left[(\overset{\langle 3\rangle}{\boldsymbol{K}}_e - \overset{\langle 3\rangle}{\boldsymbol{P}}) \cdot \mathbf{P}^{-1} \cdot \mathbf{P}^{\bullet} \right] \right]\Bigg\} \\
&= w_0(\mathbf{C}_e, \overset{\langle 3\rangle}{\boldsymbol{K}}_o)^{\bullet} + \overset{\langle 3\rangle}{\boldsymbol{S}}_p \cdot\cdot\, \mathbf{P}^{\bullet} \\
&+ \overset{\langle 3\rangle}{\boldsymbol{S}}_p \therefore \left[\overset{\langle 3\rangle}{\boldsymbol{P}}{}^{\bullet} - \mathbf{P}^{-1} \cdot \mathbf{P}^{\bullet} \cdot (\overset{\langle 3\rangle}{\boldsymbol{K}}_e - \overset{\langle 3\rangle}{\boldsymbol{P}}) + 2(\overset{\langle 3\rangle}{\boldsymbol{K}}_e - \overset{\langle 3\rangle}{\boldsymbol{P}}) \cdot \mathbf{P}^{-1} \cdot \mathbf{P}^{\bullet} \right]
\end{aligned}
\tag{6.76}
$$

with the **plastic stress tensor**

$$
\overset{\langle 2\rangle}{\boldsymbol{S}}_p := -\mathbf{P}^{-T} \cdot \mathbf{C}_e \cdot k_0(\mathbf{C}_g, \overset{\langle 3\rangle}{\boldsymbol{K}}_e) = -\mathbf{C} \cdot \overset{\langle 2\rangle}{\boldsymbol{S}} \cdot \mathbf{P}^{-T} \tag{6.77}
$$

and **plastic hyperstress tensor** defined as

$$
\overset{\langle 3\rangle}{\boldsymbol{S}}_p := -K_0(\mathbf{C}_e, \overset{\langle 3\rangle}{\boldsymbol{K}}_e) = -\mathbf{P}^{-1} * (\mathbf{P} \cdot \mathbf{P}^T \cdot \overset{\langle 3\rangle}{\boldsymbol{S}}) \tag{6.78}
$$

According to (76), the stress power goes into a change of the elastic reference energy and a dissipative part that is only active during yielding, and works on the rates $\mathbf{P}^{\bullet}$ and $\overset{\langle 3\rangle}{\boldsymbol{P}}{}^{\bullet}$.

Resumen

Our second gradient elastoplastic model is constituted by the following material laws:

the elastic reference law	$w_0(\mathbf{C}, \overset{\langle 3\rangle}{\boldsymbol{K}})$
the yield criterion (62)	$\varphi(\mathbf{P}, \overset{\langle 3\rangle}{\boldsymbol{P}}, \mathbf{C}, \overset{\langle 3\rangle}{\boldsymbol{K}}, \mathbf{H}_p)$
the flow rules (67)	$\mathbf{P}^{\bullet} = \lambda f^{\circ}(\mathbf{P}, \overset{\langle 3\rangle}{\boldsymbol{P}}, \mathbf{C}, \overset{\langle 3\rangle}{\boldsymbol{K}}, \mathbf{H}_p, \mathbf{C}^{\circ}, \overset{\langle 3\rangle}{\boldsymbol{K}}{}^{\circ})$ $\overset{\langle 3\rangle}{\boldsymbol{P}}{}^{\bullet} = \lambda F^{\circ}(\mathbf{P}, \overset{\langle 3\rangle}{\boldsymbol{P}}, \mathbf{C}, \overset{\langle 3\rangle}{\boldsymbol{K}}, \mathbf{H}_p, \mathbf{C}^{\circ}, \overset{\langle 3\rangle}{\boldsymbol{K}}{}^{\circ})$
the hardening rule	$\mathbf{H}_p^{\bullet} = \lambda h^{\circ}(\mathbf{P}, \overset{\langle 3\rangle}{\boldsymbol{P}}, \mathbf{C}, \overset{\langle 3\rangle}{\boldsymbol{K}}, \mathbf{H}_p, \mathbf{C}^{\circ}, \overset{\langle 3\rangle}{\boldsymbol{K}}{}^{\circ})$.

All other equations and conditions can be derived by these functions, like

- the isomorphy condition (52)
- the stress laws (53) and (54)
- the yield condition (64)
- the loading condition (65)
- the consistency condition (70)
- the stress power (76).

This model constitutes a rather general frame for the modelling of elastoplastic behaviour. It has been cast into a thermodynamic setting in Bertram (2016), where the consequences of the second law have been shown.

Little has been done yet to apply this general model to specific problems. Only a consistent frame has been given for this purpose, which is open for further specifications and applications.

References

Bertram, A. (2005, 2008, 2012). *Elasticity and plasticity of large deformations—An introduction*. Springer-Verlag.

Bertram, A. (2015a). Finite gradient elasticity and plasticity: A constitutive mechanical framework. *Continuum Mechanics and Thermodynamics, 27*(6), 1039–1058.

Bertram, A. (2016). Finite gradient elasticity and plasticity: A constitutive thermodynamical framework. *Continuum Mechanics and Thermodynamics, 28,* 869–883.

Bertram, A. (2015, 2016, 2017, 2019). *Compendium on gradient materials*. https://www.lkm.tu-berlin.de/menue/mitarbeiterinnen_sprechstunden/prof_dr_ing_a_bertram/.

Bertram, A. (2019). On viscous gradient fluids, *Continuum Mechanics and Thermodynamics*. https://doi.org/10.1007/s00161-019-00853-4

Cleja-Ţigoiu, S. (2002). Couple stresses and non-Riemannian plastic connection in finite elasto-plasticity. *Zeitschrift fur Angewandte Mathematik und Physik, 53,* 996–1013.

Cleja-Ţigoiu, S. (2010). Elasto-plastic materials with lattice defects modeled by second-order deformations with non-zero curvature. *International Journal of Fracture, 166,* 61–75.

Cleja-Ţigoiu, S. (2013). Non-local elasto-viscoplastic models with dislocations in finite elasto-plasticity. *Part I: Constitutive framework, Mathematics Mechanics Solids, 18*(4), 349–372.

Cross, J. J. (1973). Mixtures of fluids and isotropic solids. *Archives Mechanics, 25*(6), 1025–1039.

de Leon, M., & Epstein, M. (1996). The geometry of uniformity in second-grade elasticity. *Acta Mechanica, 114,* 217–224.

Elzanowski, M., & Epstein, M. (1992). The symmetry group of second-grade materials. *International Journal of Non-Linear Mechanics, 27*(4), 635–638.

Gurtin, M. E. (2010). A finite-deformation, gradient theory of single-crystal plasticity with free energy dependent on the accumulation of geometrically necessary dislocations. *International Journal of Plasticity, 26,* 1073–1096.

Gurtin, M. E., & Anand, L. (2005). A theory of strain-gradient plasticity for isotropic, plastically irrotational materials. Part II: Finite deformations. *International Journal of Plasticity, 21,* 2297–2318.

Gurtin, M. E., Fried, E., & Anand, L. (2009). *The mechanics and thermodynamics of continua*. Cambridge University Press.

Luscher, D. J., McDowell, D. L., & Bronkhorst, C. A. (2010). A second gradient theoretical framework for hierarchical multiscale modeling of materials. *International Journal of Plasticity, 26,* 1248–1275.

Münch, I., & Neff, P. (2018). Rotational invariance conditions in elasticity, gradient elasticity and its connection to isotropy. *Mathematics Mechanics Solids, 23*(1), 3–42.

Murdoch, A. I. (1979). Symmetry considerations for materials of second grade. *Journal of Elasticity, 9*(1), 43–50.

Reiher, J. C., & Bertram, A. (2018). Finite third-order gradient elasticity and thermoelasticity. *Journal of Elasticity, 133,* 223–252. https://doi.org/10.1007/s10659-018-9677-2.

Reiher, J. C., & Bertram, A. (2019). Finite third-order gradient elastoplasticity and thermoplasticity. *Journal of Elasticity*. https://doi.org/10.1007/s10659-019-09736-w.

Truesdell, C. A., & Noll, W. (1965). *The non-linear field theories of mechanics*. In S. Flügge (Ed.), *Handbuch der Physik*, Vol. III/3. Springer.

Index

A. Bertram and S. Forest (eds.), *Mechanics of Strain Gradient Materials*,
CISM International Centre for Mechanical Sciences 600,
https://doi.org/10.1007/978-3-030-43830-2

The manufacturer's authorised representative in the EU is Springer Nature Customer Service Centre GmbH, Europaplatz 3, 69115 Heidelberg, Germany. If you have any concerns regarding our products, please contact ProductSafety@springernature.com

Printed and bound by CPI Group (UK) Ltd, Croydon, CR0 4YY
05/10/2026
02228156-0004